THE FIELD GUIDE TO

CITIZEN SCIENCE

THE FIELD GUIDE TO
CITIZEN SCIENCE

How You Can Contribute to Scientific Research and Make a Difference

From the experts at SciStarter
DARLENE CAVALIER, CATHERINE HOFFMAN,
AND CAREN COOPER

TIMBER PRESS • PORTLAND, OREGON

"Science is a wonderful thing if one
does not have to earn a living at it."

—ALBERT EINSTEIN

Published in 2020 by Timber Press, Inc.
The Haseltine Building
133 S.W. Second Avenue, Suite 450
Portland, Oregon 97204-3527
timberpress.com

Printed in China

Text design by Anna Eshelman and Adrianna Sutton
Cover design by Julianna Johnson

ISBN 978-1-60469-847-3

Catalog records for this book are available from
the Library of Congress and the British Library.

CONTENTS

INTRODUCTION

Many times in our lives we may be filled with an urge to explore and discover. We may be curious about everyday encounters with birdsongs or spiders in webs. Or we may become concerned about air quality or the safety of our drinking water. As we face global challenges, we may want to find local ways to make a difference in protecting endangered species, safeguarding marine systems, preventing disease, or accelerating medical research. Sometimes finding solutions through new discoveries requires a lot more eyes, ears, and perspectives than scientists possess. Put simply, citizen science is a collaboration between scientists and those of us who are curious or concerned and motivated to make a difference. Citizen science can satisfy that urge, bring joy and purpose to our lives, and advance a surprising diversity of scientific research.

This book will help you discover opportunities to be an explorer, to participate in this movement sweeping the globe. Yes, the globe. If you are surprised to hear about the burgeoning popularity of citizen science, you are not alone. Conventional science frequently takes place out of sight, with methods and outcomes that remain a mystery to most. Compare that to sports, art, or music, in which we watch professionals perform in public view and then take part as amateurs in our local sports leagues, art gallery, or garage band. There's no expectation that our participation will or should lead to professional careers in pursuits we enjoy. By putting science in public view, citizen science makes it possible for anyone to participate, with or without a formal scientific background.

Citizen science brings science within reach by connecting two critical ingredients: you and teams of scientists who need and value your help for authentic research. Typically, scientists provide the instructions, protocols, and procedures, as well as the equipment and structures to guide you in sharing your observations: what you see, hear, smell, track, count, and tally. In return, you provide scientists knowledge through your observations and insights or analysis—data that scientists cannot access, collect, and analyze alone. Increasingly, citizen scientists are also setting the research agenda by identifying issues they are curious or concerned about and *then* tapping scientists to assist with the development of protocols, interpretation of data, and translation of data into action.

Today's opportunities to participate in citizen science are boundless. Odds are there is a citizen science project that coincides with any hobby, interest, curiosity, or concern that you may have. Matching people and projects appropriately is essential to success. Along with the fifty-plus projects featured in this book, we'll show you how to discover thousands of opportunities and the citizen science projects most suited to you by working with the SciStarter website. You can use it to discover, join, and even track your contributions to projects. With SciStarter as your online assistant, we encourage you to treat this book as your field guide and the beginning of an exploration into citizen science.

Some participants collect data by taking photos of clouds or streams, documenting changes in nature, or counting litter on their local beach. Other citizen scientists use low-cost sensors to help scientists keep an eye on local air, water, and social conditions. Countless others collect and send in microbes, track flu symptoms, or play games to help advance health and medical research. People just like you are counting bird, butterfly, and other pollinator populations, helping NASA track landslides, and monitoring noise and light pollution in our communities.

In short, by working together we better understand our world and make better decisions.

We sincerely hope that once you start participating in the projects featured in this book, you'll share your experiences with your friends and family and perhaps even inspire them to become citizen scientists. Witnessing the transformation in people who realize they are capable and needed gives us a sense of joy and accomplishment and inspires us to work harder to reach more people. You'll see what we mean!

THE
HISTORY
AND FUTURE
of Citizen Science

Not so long ago the idea that ordinary people possessed skills, knowledge, insights, and access to data that scientists needed raised more than a few eyebrows among the scientific elite. Frankly, even the public needed convincing that they had something valuable to contribute to the world through science. But today, citizen science is a robust, recognized, and respected field.

Along with sponsoring citizen science projects, government agencies and private and public foundations fund the work of researchers to help scientists, practitioners (project creators and researchers), facilitators, and policymakers understand best practices, learn how to design projects to improve data quality or increase participation levels and persistence, and share knowledge about the projects, people, and perspectives shaping this movement. The Crowdsourcing and Citizen Science

Act—with the goal to "encourage and increase the use of crowdsourc-ing and citizen science methods with the Federal Government"—was passed as part of the American Innovation and Competitiveness Act in early 2017. Citizen science outcomes are regularly included in peer-reviewed journals, and in 2018, the prestigious National Acad-emy of Sciences published a report titled "Designing Citizen Science to Support Science Learning." These and other developments signal a tectonic shift is on the horizon.

Much like the United States, Europe is witnessing unprecedented growth in citizen science. One way to understand the current land-scape of citizen science here and abroad is to monitor the growth in the number of projects, amount of funds available to support projects and related research, and the emergence of institutional support networks that can help the field expand and evolve. The European Union's Hori-zon 2020 initiative is largely responsible for the growth in the EU due to its substantial funding. For example, Horizon 2020 funds Doing It Together Science (DITOS), which features hundreds of citizen science events and projects embedded in art galleries, universities, science museums, and more.

Because scientific achievements have brought us to a pinnacle of health and innovation, it might seem like all research frontiers have been explored. But there will always be more unknown than known, more rocks waiting to be overturned. And discoveries are not reserved for remote reaches of the planet—they can happen close to home. In 2012, when first-year students at Columbia University were given an assignment to observe and identify ants in Manhattan, they found over a dozen species on Broadway alone. They even discovered an ant spe-cies that scientists did not know lived in New York City. What's more, that species turned out to be the most common species in the Big Apple. The reason no one had noticed? No one had ever thought to look.

CITIZEN SCIENCE: A LONG HISTORY

Over the centuries, the credentials needed to carry out scientific research have been in flux. Only recently has science become an occupation. In earlier days, science was something for those with the luxury to dedicate their leisure time or spiritual time to follow their curiosity. In the 1600s, Antonie van Leeuwenhoek discovered microorganisms. His professional background? A cloth merchant who learned to make excellent lenses to judge the weave of fabrics. Eventually, he made lenses more powerful than microscopes at that time, which allowed him to curiously examine mucky pond water and plaque on teeth and find tiny life, earning him the title of father of microbiology. Gregor Mendel filled many of his days as a monk with experimental breeding of pea plants to understand how traits are hereditary. That earned him the title of father of genetics. Charles Darwin was a companion to Captain FitzRoy of the Beagle with time to see the world before planning to return and become a parson. Darwin's later days were part of a shift in science. Not only was science becoming a profession, the precursors to citizen science was beginning: Darwin and others started crowdsourcing for data through letters in which people shared their observations from around the world.

In more recent history, fellow citizen scientists have continued to accomplish the remarkable. Citizen science has contributed hugely to entomology. The mystery of monarch butterfly migration had long eluded scientists until Fred Urquhart and Norah Patterson began experimenting with techniques to affix unique tags to butterflies. Once these scientists identified a way to attach a tag to the butterfly without harming their sensitive wings, they realized that more people were needed to help them tag as many monarchs as possible. In 1952, they asked for the help of thousands of volunteers and started a monarch tagging program, which eventually became the modern-day Monarch Watch. Then, in the mid-1970s, the first tagged monarch was spotted in Mexico. It turned out to be tagged by a Minnesota school teacher and two of his teenage students, which led to the discovery of the long-distance monarch migration from North America to Mexico in the fall and the return in the spring. The breakthrough was possible because thousands of volunteers had been capturing and tagging the wings of monarchs with postage-stamp-size stickers for decades. To this day people continue to tag monarchs and bring more discoveries, like making us aware of their current population decline.

The modern environmental movement was also inspired by citizen scientists. Rachel Carson's seminal book, *Silent Spring*, revealed the dangers of the pesticide DDT. Predatory birds, such as peregrine falcons, became endangered species because DDT thinned their eggshells. The discovery that their eggshells were thinning was possible because egg specimens found in museums had thicker eggshells. Non-professionals—citizen scientists—had collected those eggshells before the manufacturing of DDT began. (The hobby of collecting wild bird eggs was outlawed in the United States in 1916 with the Migratory Bird Treaty Act, which protected migratory birds, including their nests and eggs.)

In the mid-1990s, citizen science was key to climate change negotiations. British scientists found that birds were laying their eggs earlier in the year because of climate change. The entire dataset, with hundreds of thousands of nesting records, was the result of decades of observations by birdwatchers scattered across England. In making the case for the Kyoto Protocol (the international treaty about climate change action), the British government relied on that research to show that climate change was not a "future" problem but a "now" or urgent problem because it was already affecting life on Earth.

Today, with the internet and smartphones, science is in flux again. Millions of people, each with their own occupation (and many too young to have an occupation yet), share their observations and help process data. Volunteers work online to transcribe thousands of old letters, some originating with Darwin, others from Shakespeare, and others from war diaries. People are needed to turn handwriting into digital text because automation with optical recognition software can't decipher handwriting as well as the human eye. Today, fields like biochemistry advance because people use their free time as players in online games because the human mind is better at spatial reasoning than computers. In the Eterna game, players design RNA, the blueprints that make proteins. In Foldit, a game to solve puzzles of how proteins fold, some players discovered the folded shape of a particular protein associated with AIDS in monkeys. As environmental and health sensors like Fitbits and air-quality monitors become lower cost, people without science credentials are assessing the quality of their environment, providing a check on industries to make sure regulations are followed. In ports like Oakland, California, with significant truck traffic, and in New Orleans, Louisiana, with petrochemical refineries, communities organized by the West Oakland Environmental Indicators Project and Louisiana Bucket Brigade have discovered excessive exposures to pollution where scientists and regulatory enforcers have failed to look. Across the world, eyes of citizen scientists have discovered that endangered monk seals were attempting to recolonize the Mediterranean Sea, that invasive ladybirds in England were rapidly expanding their range, and three new species of dancing peacock spiders in Australia.

Looking across history, what's revealed is that in many areas of study the only way to keep advancing the frontiers is for scientists to collaborate, not just with each other, but with everyone.

A Commitment to Citizen Science for the Family

Exposing your children to citizen science opens them up to a lifelong interest and understanding of science and their role in science—and you'll learn a few things along the way as well. Sarah knew she wanted citizen science to be part of her family experience from before the time her twin boys were born. As she was doing the endless preparation that an expectant mother needs to do (let alone an expectant mother of twins), Sarah researched all of the potential citizen science projects she could do with her newborns. From tracking developmental benchmarks like their first time crawling to keeping a record of sleep patterns and diaper changes, Sarah was determined to make citizen science a part of her boys' experience from the start.

Sarah's recommends that parents to do their homework ahead of time if possible. The citizen science projects about newborns are easy and often things you're already tracking, but if you wait to join and read instructions after your child is born, you may struggle to keep up. If you make it part of your regular preparations you'll be ready to go once your baby arrives.

Now that Sarah's twins are a bit older, she is finding new ways to engage them in citizen science and help them build key skills. Observation using the five senses is an integral part of childhood development and citizen science teaches those observation skills that are ultimately important for critical thinking. Look for projects that give kids the opportunity to look, listen, touch, and smell. Right now, Sarah is focusing on having her kids listen to bird calls and identify what they are. By participating in citizen science projects, she's been able to see when her kids transitioned from the looking and touching development phase to beginning to listen. She might have missed this subtle transition if not for engaging her kids with science.

Sarah hopes that fostering this joy of science and citizen science early will lead to lifelong curiosity. She knows her boys will ultimately try arts, sports, music, and more, but hopes that their love of science and citizen science remains.

Ready to start introducing citizen science to your family? Sarah's advice is to seek out projects that are relatively easy and inexpensive and to be sure that any citizen science events you are considering attending have an open invitation to bring your kids along. If you're looking for something you can do on your own, she suggests iNaturalist because you only need a smartphone and you don't need much previous knowledge to get started. For slightly older children, Sarah recommends Project Budburst, Nature's Notebook, or CitSci.org, where you and your family can track changes in plants over the long term and the children can feel a sense of ownership over the project.

CITIZEN SCIENCE IS TRANSFORMING SCIENCE

In its richest form, citizen science has the power to transform science and society. Rather than simply recruiting volunteers or producing cool new tools, citizen science reshapes central notions of science and power: the roles of experts and the public, the accessibility of tools and data, and the kinds of questions that are worth asking.

As this movement continues to grow, its future isn't guaranteed to be entirely smooth or predictable. The direction it takes over the coming decades will, fittingly, be determined by developments from nearly all sectors. Government agencies will be called on to respond to the will of the people while navigating global research priorities and policies. Industry and scientists will be forced to grapple with intellectual property considerations when their core customers demand more access to data and open tools. And anyone will have the potential to shape fields of research and related policies, globally and in local communities.

Citizen science is, in many ways, a canvas with much open space remaining to be filled. And it's a movement that will undoubtedly shape your life, from finding a cure for Alzheimer's disease to gathering data about pollutants in your own backyard. Now's the time to explore the world of citizen science and look forward to the day when it is such an embedded part of our identities and lives that we no longer need any qualifier . . . it will just be called science.

YES,
You Can Be a
CITIZEN
SCIENTIST

Citizen scientists share the characteristics of being curious, concerned, and not bystanders. To have a sustainable and just world requires a new cultural norm in which being a responsible person on this planet means helping make discoveries. Observing and sharing, sometimes with smartphones and other electronic devices, is how you become part of a network taking the pulse of the planet.

As you begin your citizen scientist journey, it is common to question your own abilities and skills. Science can be complex and specialized; after all, many professionals study for a decade at universities! Skeptics and volunteers alike often wonder how citizen science can result in rigorous, reliable, and trusted science. Can strangers who may remain anonymous online and lack formal training in the sciences be authentically useful to professional scientific research projects? Can people without scientific credentials do work of sufficient quality to result in products of genuine scientific value? Is it truly possible for science-society collaborations involving individuals with variable and undocumented levels of expertise to produce reliable and trustworthy knowledge?

The answers are yes, yes, and yes. How so? As a citizen scientist, here are few things to keep in mind.

You are one of many. Citizen science is similar to democracy. In democracy, the term *citizen* refers to those with rights and responsibilities to participate in a larger collective governance. We might sign petitions, rally, and engage deeply, but the core and simplest action is that we cast a vote. Each vote may feel insignificant, but cumulatively it creates the profound impact of changing leadership. The term *citizen scientist* refers to people from anywhere in the world, living in any country, exercising their human rights and responsibilities to

What's in a Name?

Citizen science is science. Plain and simple. The term "citizen science" is one of several labels used to describe the involvement of non-professional scientists in science. This excerpt from the 2018 National Academies of Science report, *Learning Through Citizen Science: Enhancing Opportunities by Design,* sheds some light on the question, "What's in a name?"

> As a note, the committee uses the term citizen science because that is the term most commonly used within the scientific and science education communities to describe these activities. We recognize that the term "citizen," particularly in the United States, connects to a contentious immigration debate about who is eligible to participate in civic life, including science and education. While other terms can be used to describe citizen science, such as community science, public participation in scientific research, participatory action research, and community-based participatory research, none of them is as complete or widely used as citizen science. The committee uses citizen science despite its associated tensions.

participate in collective inquiry, collective discoveries. We might pose hypotheses, analyze results, and engage deeply in those ways, but the core and simplest action is that we share data. Our individual data may seem insignificant, but cumulatively it can lead to profound discoveries that can change the world. Citizen science is about the power of the crowd in which everyone does their small part, instead of relying on the heroics of an individual.

You can develop natural history expertise. Naturalists continually learn about the diversity of nature, often specializing in the details of their favorite types of animals, plants, or fungi, or developing a wealth of knowledge about rocks, stars, or weather. Natural history has always been within the purview of amateurs, not professionals. The solid foundation for all science rests on natural history observations. It is difficult for one person to be an expert in all areas. People who are not scientists have skills and expertise that are valuable to citizen science. In natural history fields in particular, expertise on species identification is held almost exclusively (other than museum curators) by amateurs. If this is not an expertise that you currently have, citizen science is one way to hone your skills. Many projects will include simple identification guides to aid in collecting data. Before you know it, you will have great natural history observation skills.

You are the expert of your surroundings. You are more likely than any professional scientist to know when something seems different. That could be a "good kind" of different (noticing fewer droughts or floods than you experienced during the past several years) or a "bad kind" (you can't use your regular bike path because of poor air quality in that area). Here's one way to think about it: If a scientist wanted to know what animals lived in backyards across an urban area, she would have to somehow visit and survey each backyard. That's not a practical approach, despite the fact that the scientist's question is valuable. Instead, citizen scientists across an urban landscape can report what lives in their backyard to the scientist, collectively answering a question that couldn't be pursued via a conventional approach.

HOW CAN SCIENTISTS TRUST CITIZEN SCIENCE DATA?

Along with citizen scientists having confidence in their own skills, citizen science projects are effective when the scientific community also understands the appropriate processes and results. For this reason, many stopgap measures are in place to reduce errors and create valuable data collected from citizen science projects. Scientists are always on the alert for data quality assurance and control. Combating errors and bias can be more challenging when more people are involved, which is amplified when projects include thousands or even tens of thousands of contributors. Fortunately, scientists have many methods to address data quality issues in project design, implementation, and analysis. Some approaches are similar in conventional science and sometimes citizen science offers new approaches for data quality. Here are some ways that scientists deal with a few common issues.

Expert review. About three-quarters of citizen science projects rely on various types of experts to validate observations. An iconic example is eBird—a project that allows you to collect, archive, and share your bird checklists—which has more than 500 volunteer expert reviewers who have reputations as excellent birders. Every year about 100 million observations are added to the open and accessible database. Every species reported to eBird is automatically cross-checked through a filtering system based on the number of birds observed, location, and date of observation. If anything exceeds what is typical for a given species at a given location on a given date, then the computer flags the observation for expert review. Reviewers then decide whether to follow up with the volunteer to request more evidence.

Photo submission. About 40 percent of projects use photographs to validate data. A community in North Carolina discovered the value of photographic evidence after the data from their community-based citizen science project was questioned. They began as volunteers combing the beach for sea turtle tracks with the Wrightsville Beach Sea Turtle Project. After volunteers encountered garbage on the beach, they began their own citizen science project, Wrightsville Beach Keep it Clean, to quantify and monitor the amount of garbage on the beach each week. When residents of Wrightsville Beach questioned the validity of volunteer findings, the volunteers modified their protocol to include photographs. Now each person brings home all the garbage they collect, rinses off the sand, sorts it into categories, and photographs it. With photo documentation, no one doubts their claims. The evidence of cigarette butts led the town council to pass an ordinance against smoking on the beach, and the overabundance of plastics led some local businesses to replace single-use plastic containers with glass containers.

Training and testing. More than 20 percent of citizen science projects have training programs related to quality assurance and quality control. For some projects, it is essential to train volunteers and/or require evidence of skills before they can make meaningful contributions. For example, volunteers for the eMammal project must complete an online training module to learn how to use motion-sensitive camera traps before they can borrow such a camera from their local library. To play the online citizen science game Foldit, which involves solving 3-D puzzles of protein folding, participants have to complete a series of tutorial puzzles, and then play games designated only for beginners until they gain the ability (as demonstrated in points) to solve harder puzzles. Hobby communities are also great for citizen science by facilitating people to teach each other needed skills. For example, for ornithology citizen science, state and local bluebird societies teach birdwatching enthusiasts to install and monitor nest boxes.

Replication by multiple participants. Almost one-quarter of citizen science projects compare data from many volunteers and validate data by independent consensus. Gathering consensus is possible for citizen science projects that rely on crowds, whether online or geographically scattered in the field. For online projects, the validation is in the form of behind-the-scenes consensus. For example, multiple volunteers must independently tag each image in Galaxy Zoo until they reach a trustworthy level of consensus, such as agreement by five, ten, or sometimes twenty volunteers. For field observations, big data allows researchers to place less emphasis on outliers (even completely eliminate them) and look for consistent patterns within the core of observations. For example, if a couple volunteers with NestWatch, a project run by the Cornell Lab of Ornithology, reported a few anomalous large clutches of bluebird eggs in late summer, that would not be enough to obscure the general seasonal pattern (larger clutches in spring, smaller clutches in summer) that is based on thousands of clutches.

Redundancy in data requests. Sometimes projects request the same data in several different ways in order to double-check for errors. For example, a project may ask for your interpretation of what you observed (e.g., when did a bird lay their first egg) and ask for details on each observation that allowed you to make that conclusion (e.g., what you saw during each visit to a bird nest). Another redundancy to help with double-checking for errors is requiring data entry on paper forms as well as through online entry.

A good match. A project requiring participants with particular expertise will fail if the participants are poorly matched to the project. Here's an example of a poor match from the early days of citizen science: In the late 1800s, the newly formed American Ornithologists' Union recruited lighthouse keepers to report on the fatalities of migrating birds hitting the lighthouses. Unfortunately, most lighthouse keepers were not birdwatchers and were unfamiliar with the taxonomy and nomenclature of bird species. Their observations of colloquial bird names like sea robins, Mother Carey's chickens, black sea ducks, and bee martins were indecipherable to ornithologists who did not know how to translate those names into species. On the other hand, when entomologists needed help finding the nine-spotted ladybug, the Lost Ladybug Project recognized that kids were ideally suited to the task of what became equivalent to a massive treasure hunt. Ultimately, populations of nine-spotted ladybugs have been found by kids.

Prepare to
Take the Pulse of
THE PLANET

Achieving a great  citizen science experience is more likely if your personal goals and expectations of the project are clear. Are you looking to socialize and join forces with a community of citizen scientists? Would you prefer to do that in person or online? Or would you rather work autonomously? Are you comfortable with (or do you want to learn how to build or use) instruments to collect or analyze data? How often can you participate? Every day? Maybe every month? Or just once a year? Do you want to learn a new skill or use existing experiences? Are you passionate about a particular topic or issue? Would you like to contribute to a local, regional, national, or global project? These are all important questions to ask yourself before getting started. If you are going to invest time and energy into a citizen science project you should do everything possible to get the most out of it.

TIPS FOR A SUCCESSFUL COLLABORATION

When selecting a citizen science project, consider if the project has clearly stated what they are expecting from you as well as what you can expect from them. Even though citizen science has a long history, its formalization is new. Universities are beginning to graduate public scientists, but training for scientists in the techniques and best practices of citizen science or public engagement is still new and continuing to evolve. The management of citizen science projects frequently involves teams of people from a range of disciplines, including science education, communication, information sciences, human-computer interaction, and more. This makes for a dynamic and efficient team, but it also means that individual players have strengths and weaknesses. When it comes to citizen science, *you* are the expert, so it's important to make sure your voice and feedback is heard.

Here are four questions to contemplate when choosing a project:

1. **Are you on the same page?** Frequency of expected participation is an important consideration in this regard. If the project is expecting weekly contributions, the scientists might be frustrated when you only participate once. Conversely, if you participate frequently in a project that only responds to participants monthly, you might feel that same frustration.

2. **Will you be treated like a member of the team?** Scientists should treat you as a respected member of the research team, irrespective of whether you are joining a project with ten other volunteers or tens of thousands of other volunteers. If you want to engage more

deeply than the prescribed protocol, the scientists should be receptive to your input. But remember: not every project has the capacity to support deeper-than-average involvement, so you'll have to find the project that best fits your interest and commitment level.

3. **Will your work be valuable?** People volunteering their time do not want to simply feel like they are helpful, they want to actually be helpful. Scientists should never assign work that is not authentically useful. For example, they should not ask you to do tasks that a computer algorithm could complete, just for the sake of engagement. (However, be aware that computers cannot do many tasks that are easy for people, particularly when it comes to image recognition and solving puzzles.) Additionally, the sole aim of a project should not be to teach you something. Good citizen science elicits knowledge from citizen scientists to advance scientific research or to solve a real problem.

4. **Will you receive feedback?** All citizen science projects should share updates and results with their volunteers. While some projects must keep some (or all) data private, there's really no excuse for scientists not to provide you with updates. Other than a lack of time, the number one reason people stop participating in a project is when they don't hear back from the project leader. Of course, science is a process so there just might not be anything to report between when you submit data and the time it takes for the data to be analyzed and interpreted. But, at the very least, you should receive a message or email to confirm you successfully submitted data.

Do keep in mind that citizen science projects commonly operate on shoestring budgets. (Because citizen science relies so heavily on volunteers, organizations that fund scientific research often underestimate what's required to successfully run a citizen science project.) This means that the organizers may be underprepared for the amount of involvement required—especially when the project is launching. It's important to understand the limits put on project managers so you can manage your expectations. However if you don't receive some level of feedback you should consider trying a different project, or at least reach out to the project manager to find out whether the lack of communication is a choice or a constraint. If you volunteer to do more for the project (perhaps assist with communications such as newsletters, blogs, social media, emails, or forums) they may take you up on that.

FOLLOW THE PROJECT PROTOCOLS

Whether you've chosen an online project that requires you to mark and transcribe images of lunar surfaces or you will be monitoring local bodies of water, every good citizen science project should have clear protocols and methods to follow. Sometimes these guidelines can seem overwhelming, but they are important to ensure quality data. Citizen science is serious science and we want the data you collect to be taken seriously.

Whenever scientists investigate something, they spend time thinking about how to collect the right data and how to create protocols. Let's say scientists were trying to understand why an animal lives in a certain location. They would need to design a study that carefully examined the influence of environmental factors, like rain, or the presence of other living things, like predators, on that species. They would need to figure out how to count the animals in the best way, how to measure the rain and the predators, and make sure that everyone on their team was using the same protocols to collect data in ways that could be repeated over and over. Now imagine setting up a citizen science project for hundreds or thousands of people to collect data in the same way. Project organizers need to be especially careful about what they're asking citizen scientists to do.

Just as Julia Child's primary cooking advice was to read a recipe entirely before starting, take some time to review all the instructions, training modules, and information about your citizen science project before you begin. If you have questions, get in touch with the project manager, ask other citizen scientists, or consider forming your own club or group to approach citizen science with you. You'll want to be prepared before you get started. When you are prepared, you will be able to collect important data for the project.

HOW TO MAKE
THE MOST IMPACT

Citizen science involves observing and then sharing what you observe. In some cases, all a scientist may need is for you to share your haphazard observations. More commonly, for your observations to be of greater value and increase the likelihood of being useful for scientific discovery, scientists need you to slightly tweak how you make observations.

For example, birdwatchers used to keep lists that simply included the species observed in a given place at a given time. That alone is tremendously useful, but citizen science birdwatchers have refined the process to make it even more useful. The standard is now to keep lists of how many individuals of each species were observed in a given place at a given time, and to note whether the list includes all species observed (as opposed to just the birdwatcher's favorites, or just the songbirds, or just the native species, etc.), and to note the birdwatcher's effort (how long they were out with the intention of observing birds).

The same goes for projects that involve collecting samples. Let's say you wanted to sample for lead in your water. You might think you can just use a plastic water bottle, fill it with your tap water, and send it off for sampling. But scientists may need you to use a sterilized bottle, let the water sit for eight hours in your pipes, or even follow a chain of custody paperwork to ensure that no one else can tamper with the sample. The project protocols likely describe all the required steps so remember to read carefully and consider the data you might otherwise not think to collect.

Let's break this down further with some helpful tips:

Record and share what you do *not* see. The absence of a species, event, or phenomenon can be as important as its presence. Have you always seen migrating butterflies in your yard each April, but this year you saw none? That's important. Perhaps your observation of no butterflies indicated a shift in the range of where they can live. Other notes about the absence of particular species may indicate the local extinction of a species.

Record and share what's common and not just what's rare. Sometimes we ignore what's most common around us, especially the good things. When it comes to biodiversity, sometimes the most common animals can suddenly decline. When it comes to pollution, sometimes air quality can abruptly plummet. If you haven't been documenting the many ladybugs in your garden or the many days with clear air, it will be harder to prove a sudden degradation in environmental quality.

Record your effort. In many cases, scientists need to know how much time you spent on the alert for an observation or how long you spent collecting data. For example, did you count the birds in your yard for one minute or ten? That's going to affect how many birds you're likely to see and what that means for the research questions. Or perhaps you collected a water sample from a stream, which only took one minute, but you hiked for two hours to find the remote stream. That might be important for the scientist to know.

Participate during the weekdays too. Most of our free time is on the weekend, so that's when we are most likely to do citizen science. Weekend bias is a particular form of bias common to citizen science, and once identified, scientists can correct for it statistically. But you can also correct for it methodologically by taking some free time for citizen science during weekdays too. So make sure you observe your bird feeder in the mornings before work as well as during a lazy Sunday morning.

Ask questions. Most scientists would prefer you to ask for clarifications and would appreciate your input on how to make instructions clearer.

READ THE FINE PRINT

Each project should have a terms of service, informed consent, privacy policy, or all three. These are important for several reasons. First, they show that the project has gone through the proper channels to collect and manage data. If a project is collecting personal data you should expect to see an informed consent. This typically means the project has gone through an institutional review board to conduct studies that collect data about people. Privacy policies are important so you know how your information and the data you share are used. Don't be afraid to ask if you are unable to find these documents or have questions regarding your privacy.

Scientists are the custodians of citizen science data. Once you share data with a project, you will want to know where it went. Your submitted data will be treated differently depending on the project, somewhere on the spectrum between fully available (open) data and fully private data. If having ownership of your data and/or knowing that it is publicly available is important to you, then be sure to investigate data usage policies before you join a project. Many projects explain this in the project description, privacy policy, or terms of use.

Open vs. Private Data

Open data means you will have access to the data you contribute to the project as well as data submitted by everyone else. You'll be able to see your data on a map, in a spreadsheet, or through an online interface, often right away. You can also look at data from other people in your community or around the world, and sometimes you can even download the data. Examples of projects and platforms that post all the data and individual data points include AirCasting, eBird, Globe at Night, iNaturalist, ISeeChange, Monarch Watch, Project Squirrel, Stream Selfie, and ZomBee Watch.

Private data may mean that you can only view the data you shared. You may or may not be able to view other data. But just because a project keeps its data private, doesn't mean the scientist has bad intentions. Scientists may keep data private for several reasons:

* The data might contain personal or confidential information. For projects in the biomedical field, for example, you might submit information about an autoimmune disease, side effects of endometriosis, or signs of dementia. This information is important, but the researchers must keep it private or anonymous.

* The data could include sensitive information. If the data you share could cause harm to you, someone else, or even a species, the scientist might keep the data private. For example, when someone submits an observation of a threatened or endangered species on iNaturalist, the site does not allow the precise location to be shared publicly to protect that animal or plant.

✳ Scientists can be competitive. This isn't a noble reason, but it is a common concern among those making a career in the sciences. Professional advancement is based on merit, which is judged according to discoveries made. Scientists often keep data private in order to have time to be the first to make discoveries from it.

Many projects fall somewhere between fully open and completely private. Maybe the data is private until it has gone through a review process or maybe the project organizers only show a subset of observations on a map. Again, read closely and ask questions if something isn't clear.

Get Comfortable with SciStarter

SciStarter is a research affiliate of the School for the Future of Innovation in Society at Arizona State University and widely used by North Carolina State University to study the field of citizen science. SciStarter is a popular citizen science portal with an active community of citizen scientists and project leaders. Thanks to support from the National Science Foundation, SciStarter brings citizen science projects together in one central place, helping you track your interest in and contributions to different projects. More than three thousand projects and events representing a wide array of topics have been registered by individual project leaders or imported through partnerships with federal governments, NGOs, and universities.

The SciStarter team is working to continually improve SciStarter with tools to help people navigate the world of citizen science and find their best place in it. As the world of citizen science evolves, SciStarter will be there to help you stay on the cutting edge. By becoming familiar with SciStarter and its resources, you will be able to find the best citizen science projects and discover a community of people who share your desire to act.

Here are a few ways that you can use SciStarter to improve your citizen science project experience:

- Create an account on SciStarter.org. A SciStarter account helps you track your contributions and find relevant projects to participate in. Completing your profile is also important so scientists who need you can find you.

- Use your SciStarter dashboard to find recommended projects and events, view your portfolio of project contributions, view and join bookmarked projects, and more.

- To find a project that is happening near you (or globally), enter your location in SciStarter's project finder tab and click "find projects." You can also use this feature to narrow down projects to match your interests, skills, and more.

- If you participate in a project and never hear back from the scientists, let other would-be participants know when you rate and review the project on SciStarter.

- Reach out to the project organizer by clicking the "message project leader" button, which is available on any project page on SciStarter.

- When you join projects, be sure to use the same email address you used to sign up for SciStarter. This way, you can earn credit for your contributions on your SciStarter dashboard.

CITIZEN SCIENCE

Projects to do Online, in Your Home, or in Nature

Where can you do citizen science and who can do it? The easy answer, of course, is everywhere and everyone. In this chapter, you will find projects that are open to participants across the globe (including those you can do entirely online) as well as projects that are limited to specific continents, countries, states, or regions, such as the coastline of California. Check the location line of each project to make sure you choose a project that's the right fit for where you live and explore. And remember, you can find thousands of additional projects through SciStarter too.

Projects typically take place wherever observations and data are needed. Project Squirrel, for example, invites citizen scientists to share observations of where they see different types of squirrels, what the squirrels are eating, and what their habitats are like. This project is open to anywhere in the world where squirrels are present. Stream Selfie—which encourages participants to take pictures of local streams, verify whether or not the streams are included on a national map and share data about the health of the stream—takes place anywhere in the United States where there are streams.

Many projects—from classifying clouds to identifying insects—can be done anywhere on earth. Projects with a global reach can engage thousands, if not millions, of people in a common goal to study science. Many first-time citizen scientists participate to make an impact on something they care about, whether it's a new or ongoing environmental issue or a medical problem impacting themselves or their loved ones. So while global projects are accessible in scope, make sure to ask yourself, "Why does this matter to me?"

If you are looking for a citizen science project to do with your local youth organization, scout group, or community center, we recommend considering how it connects to your community. Talk to the people who you want to participate in the project. What do they care about? What types of questions do they want to help answer? Is there a project that fits these needs? For example, a local park community might want to track the biodiversity found across the park so they can create successful wildlife management plans. Or, a school might want to track rainfall on campus to know when to water their garden. There are many projects and platforms to help address global or national questions through a localized lens.

This chapter is filled with projects that will complement your current hobbies, activities, and interests, as well as plenty of opportunities to branch out into unknown territory—even the extraterrestrial. The projects are divided into the following categories: online, at home, near water, at night, out and about in your community, and chance participation. Unless special materials are listed, you will generally only need a computer, tablet, or smartphone with internet access (and a camera if you don't have a phone with picture-taking abilities) to participate in most of these projects. Also note that we selected these particular projects because they represent a diversity of activities and topics, they are registered on SciStarter, they are likely to be active for years, and they advance scientific research. Just remember: they are the tip of the citizen science iceberg.

SciStarter Affiliate Projects

Many projects on SciStarter are labeled as "affiliate projects," which means that if you have created a SciStarter account, you will earn credit for your contributions to these projects. If you need evidence of your participation in citizen science for your school, troop, or volunteer service hours, for example, simply print or share your SciStarter dashboard. The Girl Scouts of USA, some school districts, universities, and companies assign members affiliate projects through SciStarter's customized portals. The portals provide these member organizations with data analytics to measure collective impact and help ensure you get credit for your contributions. Here are the affiliate projects included in this book:

- Ant Picnic
- Backyard Bark Beetles
- Budburst
- CoCoRaHS
- Crowd the Tap
- DeepMoji
- Dragonfly Swarm Project
- Flu Near You
- Globe at Night
- iNaturalist
- ISeeChange

- Marine Debris Tracker
- Nature's Notebook
- Project Squirrel
- Stall Catchers
- Stream Selfie
- The Great Pumpkin Project
- TreeSnap
- Wet-Dry Mapping
- ZomBee Watch

ONLINE

As an armchair citizen scientist, you can advance science while retaining the flexibility to participate when and where you want. You may help classify images of stalled blood vessels, advance the emotional capability of artificial intelligence, or transcribe museum labels from the comfort of your home or wherever else you and your device may be passing time. Online-only projects don't typically offer the chance to meet other participants in person—although a select few do—but many platforms have active community forums, blogs, or social media groups to ask questions, discuss the project, and interact with other citizen scientists.

Bat Detective

Feeling a bit batty? There's a citizen science project for that. These important nighttime animals are threatened by changing habitats and disease. Bats are also hard to find and identify, so scientists rely on citizen scientists to help identify bat calls with the ultimate goal of understanding how global change is affecting bat populations. No experience is needed to identify bat calls online through Bat Detective.

Location Global

Website Search "Bat Detective" on SciStarter or visit batdetective.org

Goal To collect data on bats around the world

Task Complete a brief tutorial on the project website about how to analyze a spectrogram and identify the noises that you hear. For each spectrogram, you will play the audio clip and mark the frequency of the sounds, highlight the individual noises, and use the field guide to identify the sound.

Outcomes Data analysis from Bat Detective is helping scientists create and improve upon automated bat call detection tools.

Why we like this Bats are important indicators of healthy ecosystems around the world, but their nighttime flight means they're harder for the general public to understand. Bat Detective makes it easy for anyone to help in bat research and appreciate their value.

Beats Per Life

"Lub dub, lub dub, lub dub." Your heart has a distinctive beat that you carry with you through life. Have you ever wondered how many beats you get? Studies have shown that mammals get one billion beats in their entire lifetime. However, some animals may use all those beats quickly in their short life, while others live for a long time and have the same number of beats. In this project, you'll help scientists from North Carolina State University understand the heart rates of various animals to ultimately help unlock new secrets to long lives.

Location Global

Website Search "Beats Per Life" on SciStarter or visit robdunnlab.com/projects/beats-per-life/

Goal To learn more about animal heartbeats

Task Start by looking at the animal heartbeat data that citizen scientists have already collected and choose an animal that is missing data. Look online and in scientific literature (find tips and guidance on the project website) to find the heartbeat data. Once you find the data for your chosen animal, enter it in the project's online form.

Outcomes As of 2019, citizen scientists have contributed heartbeat data for 176 mammals, 60 birds, 45 reptiles or amphibians, and 41 fish.

Why we like this You may have never thought your internet sleuthing skills could be used for science, but this project shows that they can. Even though you're just tracking down data, you'll certainly learn a lot along the way.

C-BARQ and Fe-BARQ

Are you a dog or a cat person? Or maybe you love both. No matter what, C-BARQ and Fe-BARQ—Canine (or Feline) Behavioral Assessment and Research Questionnaire—have you covered. To participate in this citizen science project from the University of Pennsylvania you'll need to know a dog or cat well enough to complete a fifteen- to twenty-minute survey about your pet's daily behavior. Your data will help researchers understand the variations in behavior and temperament in different dogs and cats.

Location Global

Website Search "C-BARQ and Fe-BARQ" on SciStarter or visit vetapps.vet.upenn.edu/cbarq/

Goal To create standardized evaluations of dog and cat temperament and behavior

Task Add each pet you want to evaluate to your online portal on the project website. You'll need to know information about your pet's age, breed, gender, weight, and behavior. Complete the short evaluation survey for each animal you add.

Outcomes The project has collected data on more than 50,000 dogs and about 5,000 cats. The data is available to adoption centers, veterinarians, breeders, trainers, and more.

Why we like this C-BARQ and Fe-BARQ take information you probably already know about your four-legged friends and allow you to share that data for science. The survey questions are easy for any pet owner to answer. Plus, it might help you think a little more deeply about your pet's behaviors.

DeepMoji

In our digital world, we've all written something online and wondered, "Will they understand what I mean by that?" It can be hard to always know the inflections and mannerism associated with written text. But for artificial intelligence (AI) systems it's even more difficult. DeepMoji is a MIT project that is helping to advance psychologists' understanding of emotions. In the project, you will be asked to review your last three tweets and indicate what your feelings were when you sent the tweets. The data is then converted into emojis, one of the ways we commonly express our emotions. The project helps uncover the nuances in human language that might be otherwise overlooked in AI. Researchers ultimately hope to improve the ability of AI to identify bullying and racism online.

Location Global

Materials A Twitter account

Website Search "DeepMoji" on SciStarter or visit deepmoji.mit.edu

Goal To teach AI about emotions and advance emotion research

Task The project website will prompt you to sign into your Twitter account and your last three tweets will be displayed. Then you will be asked to recall your emotions at the exact moment that you wrote those tweets. It should only take about five minutes to complete the questions.

Outcomes DeepMoji continues to process tweets and improve its AI algorithm. The data from citizen scientists is increasing understanding about emotion and sarcasm, and more progress and breakthroughs are yet to come.

Why we like this Who knew our tweets could be so important for science? But once you evaluate your tweets with DeepMoji, you'll realize how hard it is for AI to interpret the complex emotions within those simple sentences.

Flu Near You

Help prevent the next epidemic by taking a few moments each week to report your flu symptoms (or lack thereof) to epidemiologists at Harvard University, Boston Children's Hospital, and the Skoll Global Threats Fund. They will analyze and anonymously map the information to provide public health officials with real-time, anonymous data. Note: If you are experiencing flu symptoms, always share them with your doctor too—this project is not a substitute for medical care.

Location Global

Website Search "Flu Near You" on SciStarter or visit flunearyou.org

Goal To predict and prevent the next flu pandemic

Task Create an account on the project website or download the Flu Near You app, if desired. Spend a few seconds each week to answer the brief survey questions, whether or not you have flu symptoms. View the data map and check back in again the following week.

Outcomes Thousands of participants have tracked their flu-like symptoms with Flu Near You. The project's novel crowd-sourced method of tracking flu outbreaks has also led to multiple peer-reviewed scientific publications about the project.

Why we like this According to the U.S. Centers for Disease Control and Prevention, the flu kills approximately 80,000 Americans each winter. Along with the flu vaccine, understanding and predicting outbreaks is essential to combating this killer. Flu Near You enables everyone to help produce credible data to predict local, state, and national flu epidemics.

Foldit

Knowing the structure of a protein is key to understanding how it works and to targeting it with drugs. A small protein can consist of 100 amino acids, while some human proteins can be huge (1,000 amino acids). The number of ways even a small protein can fold is astronomical so figuring out the best possible structure is one of the hardest problems in biology. Current methods take a lot of money and time, even for computers.

Foldit is a revolutionary computer game enabling you to contribute to important scientific research. Foldit attempts to predict the structure of a protein by taking advantage of humans' puzzle-solving intuitions and having people play competitively to fold the best proteins. This project is collecting data to find out if our pattern-recognition and puzzle-solving abilities make us more efficient than existing computer programs at pattern-folding tasks. If this turns out to be true, we can then teach human strategies to computers and fold proteins faster than ever.

Location Global

Website Search "Foldit" on SciStarter or visit fold.it

Goal To help discover and predict the structure of a protein

Task Create an account on the project website and download the game. Begin playing the introductory training puzzles using the forums for help. Persist through the required puzzles, join a team, and begin solving new puzzles.

Outcomes Protein folding is key to several diseases including HIV/AIDS, cancer, and Alzheimer's. The contributions of Foldit teams are opening the door for new advances to fight these diseases.

Why we like this Teams that successfully solve protein-folding puzzles become coauthors on scientific papers. Plus, teams become close-knit partners, helping each other help science.

Galaxy Zoo

Galaxy Zoo needs your help classifying galaxies according to their shapes—a task your brain is better at than even the fastest computer—in order to understand how these galaxies, and our own, formed. The project's files contain almost a quarter of a million galaxies that have been imaged with a camera attached to a robotic telescope (the Sloan Digital Sky Survey).

Location Global

Website Search "Galaxy Zoo" on SciStarter or visit galaxyzoo.org

Goal To map the galaxies by viewing telescope images online

Task Classify galaxies according to their shape and help scientists determine how the shape of a galaxy affects the light spectra.

Outcomes More than 200,000 people have already taken part in Galaxy Zoo, producing a wealth of valuable data and sending telescopes chasing after their discoveries.

Why we like this Yours might be the first eye to actually see the galaxies in this project because a robotic telescope collected these images rather than a human observer.

SETI@home

Does intelligent life exist beyond Earth? The search for extraterrestrial intelligence (SETI) can be done from the comfort of home and requires minimal effort with this project. By downloading the free BOINC screensaver, you can connect your computer to other internet-connected distributed computers from around the world to analyze and report radio telescope data to SETI scientists at the University of California, Berkeley.

Location Global

Website Search "SETI@home" on SciStarter or visit setiathome. berkeley.edu

Goal To search for intelligent life in space

Task Visit the project website and download the BOINC screensaver onto your desktop or laptop computer. Your donated computer processing power will help aid scientific supercomputers in their search to detect and analyze potential extraterrestrial signals. Your computer's processing power will only be used when your computer is idle (in other words, when you're not using it). Check back often for new developments.

Outcomes SETI@home has demonstrated that millions of people are willing and able to advance scientific research through distributed computing power. Although no specific radio signals have detected signs of intelligent life in space, the massive data analysis has helped scientists zero in on several hotspots.

Why we like this It doesn't get much easier than this project. And, really, aren't you curious to know if intelligent life exists beyond Earth? We sure are.

Smithsonian Transcription Center

Historical documents in museum collections across the world are often inaccessible despite our ever-growing digital infrastructure. You can help unlock historical documents and biodiversity data by becoming a digital volunteer with the Smithsonian Transcription Center. Volunteers help increase accessibility of Smithsonian collections by transcribing handwritten documents and specimen labels. After the transcription is completed and reviewed by Smithsonian staff, it will become available to be downloaded and the transcribed text will also become searchable through the Smithsonian's online database and other major search engines. This work improves collection records and creates greater access to the multitude of information within these materials.

Location Global

Website Search "Smithsonian Transcription Center" on SciStarter or visit transcription.si.edu

Goal To make historical data more accessible for research and discovery

Task Create an account on the project website and review the tips page for general and project-specific instructions. (Volunteers can also contribute anonymously but registering allows you to review the work of others and edit where necessary, keep track of your progress, and receive a monthly newsletter.) Browse projects on the website to find something you are interested in by Smithsonian unit or theme. You don't need to have any prior knowledge of the topic, just a commitment to transcribe as accurately and completely as possible.

Outcomes Since the center's creation in 2013 more than 380,000 pages of field notes, diaries, ledgers, logbooks, currency proof sheets, photo albums, manuscripts, and biodiversity specimen labels have been transcribed and reviewed. The amazing community has grown to over 11,600 digital participants.

A Mom Finds Her Calling as a Volunpeer

Seeking an easy way to participate in work that was stimulating yet flexible enough to fit into her busy schedule, Siobhan got started in online citizen science. She liked that the format was educational and challenging but didn't require a change in her schedule. She found the online volunpeer community at the Smithsonian Transcription Center to be supportive and collaborative. Her first project involved transcribing the field journal of an early-twentieth-century naturalist. Once transcribed, the biodiversity data collected in the journal could be added to an easily searchable and usable form. From there, Siobhan transcribed specimen labels on plants and insects that scientists could then search and analyze.

When choosing a project to do online, Siobhan recommends finding out how you'll be able to interact with other participants and project leaders. Siobhan hangs out on Twitter and she'll often look up project handles or hashtags to see how active they are. The active community isn't necessarily what makes her pick a project, but it is definitely what makes her want to stay.

Siobhan's experience with online citizen science got her started in more biodiversity and conservation efforts in her hometown. She takes part in projects to eliminate invasive species and provide data to manage pest-free regions of New Zealand. She's also used her experience in citizen science to better plan a garden full of native species. Once her garden is planted, Siobhan tracks the flora and fauna it attracts in the iNaturalist app.

Siobhan encourages new citizen scientists to try a variety of projects to find what really interests them. Reach out to the project manager and fellow volunteers whenever possible. Digital volunteering may seem isolating, but Siobhan has made a close group of friends in the online community. "Behind every project," she advises, "are like-minded enthusiasts who are keen to be contacted about the content, issues, and challenges of the project." So what are you waiting for?

Why we like this This project is an opportunity for the public to engage with the Smithsonian's collections and improve their materials. Every contribution helps, no matter how big or small. Plus, the robust online community—the "volunpeers"—is one of the best out there. This project is great if you're looking for an opportunity you can do completely online, but also want the community feeling of an on-the-ground citizen science project.

Stall Catchers

Across the United States, 5.7 million people are living with Alzheimer's disease, the seventh leading cause of death in America, yet there is no treatment or cure. It hits close to home for many of us who have seen loved ones suffer and who feel hopeless in the face of this disease. With Stall Catchers, joining the fight against Alzheimer's is as easy as playing an online computer game.

Recently, scientists at Cornell University found a link between "stalled" blood vessels in the brain and the symptoms of Alzheimer's. These stalled vessels limit blood flow to the brain by up to 30 percent. In experiments with laboratory mice, when the blood cells causing the stalls were removed, the mice performed better on memory tests. Therefore, scientists are working hard to develop treatments that remove the stalls in mice with hopes of applying their understanding to humans. But analyzing the brain images to find the stalled capillaries is hard and time consuming: it could take a trained laboratory technician six to twelve months to analyze each week's worth of data collection. Even though their findings were promising, it could take decades to run the series of studies needed to arrive at a treatment target because of this analytic bottleneck. So Stall Catchers was created to make finding the stalled blood vessels into a game that anyone can play. The game relies on the power of the crowd—multiple confirmed answers—before determining whether a vessel is stalled or flowing.

Location Global

Website Search "Stall Catchers, by EyesOnALZ" on SciStarter.org or visit stallcatchers.com

Goal To speed up Alzheimer's research

Task Create an account on the project website and start playing the game to report stalled, clogged blood vessels in moving images of mouse brains. You will receive tips and tutorials as you walk through catching your first stalls. Play the game often to continue catching stalls and improving your skills.

Outcomes In the first month after launch, 1,000 users analyzed 96,000 blood vessels, producing crowd answers that achieved over 95 percent accuracy, and in some cases even revealed mistakes that experts had made. Today, more than 10,000 people are helping accelerate Alzheimer's research through Stall Catchers. The team is currently inserting into Stall Catchers the first dataset that tests a prospective treatment mechanism based on disrupting the formation of stalls.

Why we like this Stall Catchers is a win-win project that combines scientific research into a widely accessible game. Because Alzheimer's is close to many of our homes and hearts, accelerating progress is even more meaningful. You will also find a community of "catchers" around the world with whom you can interact, learn from as you start the project, and compete with for points in the game. The team behind Stall Catchers sometimes hosts in-person "catchathons" to spark competition in the game over a weekend.

A Citizen Science Medical Quest

Vivek is not your typical citizen scientist. He didn't get involved to enjoy the outdoors or help scientists with a problem or learn a new skill. He got involved to help himself. Vivek has an autoimmune disease that has caused myriad symptoms over his life. Some persistent, others changing. Despite his diagnosis and input from dozens of doctors, Vivek wasn't feeling any better. As he searched for answers to his questions online, he realized there were hundreds if not thousands of forums, blogs, websites, and more dedicated to understanding symptoms, causes, and solutions for those with autoimmune diseases. It wasn't just Vivek who was unsatisfied with his care—countless others were struggling to find their own answers too. So Vivek got to work.

As a software engineer, he could see the potential for uniting this information to create a robust set of data that could be analyzed and shared among the autoimmune community. In 2015, Vivek launched Autoimmune Citizen Science (AICS) to help pull together the data that was spread across forums into one place. Individuals can track their symptoms, lab tests, diets, medications, and more to allow for individual analysis. AICS participants are already seeing results. One individual found she was better equipped to manage her own care after determining that her body temperature was not only associated with her menstrual cycle, but with another symptom she didn't realize was connected.

The real power of the AICS tracking app comes into play when the data is aggregated and shared anonymously. The work of individual citizen scientists investigating their symptoms creates a strong force to examine correlations, understand new treatments, and, potentially, drive life-changing autoimmune research. With health and medicine citizen science you have the unique opportunity to help others, and to ultimately help yourself.

You can find lots of citizen science projects that deal with personal health and medicine on SciStarter. Many of the projects, like AICS, are looking for people with a certain condition. However, other projects also need control participants— those that are not experiencing issues—to help validate their results. Citizen Endo, for example, asks for participants who have endometriosis, and for those that do not, to track their symptoms.

Ready to get started? Here's Vivek's advice: "You're going to get excited and want to track tons of symptoms, activities, medications, and body measurements. But focus on a few that are important to you so you can stay consistent when reporting results."

AT HOME

_ . _ • _ ● _ • _ . _

Your home and backyard can become your own scientific research lab with citizen science. Scientists have collaborated with citizen scientists to learn about everything from the microbes in our showerheads, the quality of our drinking water, to the soil in our backyard. Get ready to step away from the computer, get your hands dirty, and make some discoveries—all without venturing too far afield.

Backyard Bark Beetles

Tiny bark beetles have destroyed approximately 46 million of the 850 million acres of forested land in the United States. The Backyard Bark Beetles project is working to identify where these beetles are found across the country. Some bark beetles actually help forests by clearing dead wood, but many are destructive to forests as well as to agricultural fruit trees. With this project, you'll create a homemade trap for beetles in your yard and send the collected beetles to scientists for species identification. This low-cost project is ideal for schools, community groups, and scouts.

Location United States

Materials Trap-making supplies (two-liter bottle, string, alcohol-based hand sanitizer), spoon, Ziploc bags for samples

Website Search "Backyard Bark Beetles" on SciStarter or visit backyardbarkbeetles.org

Goal To monitor bark and ambrosia beetles across the United States

Task Follow the instructions on the project website to create a simple DIY bark beetle trap with a two-liter bottle, string, and hand sanitizer. Hang your trap from a tree in your yard. The hand sanitizer at the bottom of the bottle will help bring the beetles to your trap (many are attracted to the scent of ethanol) and preserve the beetles you find.

Crowd the Tap

Lead pipes are the primary way that lead enters drinking water. Crowd the Tap is an EPA-funded project that promotes access to safe drinking water by using citizen science to identify the materials of pipes that deliver drinking water to homes. Although installation of lead pipes was banned in 1986, and some water systems and homes replaced lead pipes, much of the drinking water infrastructure in the United States likely still contains lead pipes. Citizen scientists are needed to create a robust national inventory of water pipe materials as currently there is only piecemeal data on which water systems still have lead pipes. The inventory is used to identify areas with the highest risk of lead exposure in order to direct triage efforts. Crowd the Tap hopes to connect people and empower residents to take an active role in making sure their tap water is safe.

Location United States

Materials A penny and a small magnet

Website Search "Crowd the Tap" on SciStarter or visit crowdthetap.org

Goal To inventory tap water pipe materials used in plumbing, service lines, and water mains

Task Inspect and identify the types of plumbing and service line where you live and contact your water provider to determine the type of water main. The service line that brings water into a home is buried underground and so volunteers have the best place of viewing where it connects to the water meter. Water meters can be in the cellar, basement, or garage, or even outside. Volunteers use a magnet and a penny to figure out the materials of the service line entering the water meter as well as their home plumbing as it leaves the water meter. If a magnet sticks to a pipe, then the pipe is galvanized steel. If a magnet does not stick, then volunteers scratch the pipe with a penny to distinguish whether the pipe is plastic (no shine when scratched), copper (shines

At Home . . . With Microbes!

Nothing beats a cozy day at home by yourself. But are you ever truly alone? Supernatural phenomenon aside, citizen science projects through Rob Dunn's lab out of North Carolina State University have shown that thousands of species also call your house "home," including visible intruders like cellar spiders or crickets as well as microbes living across all of your surfaces.

A hot shower after a long day is rejuvenating, but your showerhead can be home to an invisible world of microbes. Citizen scientists were asked to swab their showerheads across the world and submit the samples to Dunn's lab for analysis, which helped unveil the unique communities of bacteria found just above our heads. Many of these microbes are completely harmless, but some have been linked to disease.

It's not only the microbes in your showerhead, but the microbes in your armpits that are your co-residents. Glands in your armpits actually feed microbial bacteria on your body. We still don't know much about them, but we do know that those bacteria show up around our houses. How are we learning about armpits? You guessed it . . . citizen scientists swabbed their armpits for science.

And finally, don't forget the furry friends that call your house a home. They've got their own world of microbes that they're bringing inside to join you each day.

the color of the penny when scratched), or lead (very shiny silver when scratched). The service line and home plumbing is only half of the water system. Some volunteers contact their water provider and find out the material of the water, service lines on public property, and the gooseneck (the pipe that connects the water main to service lines). Volunteers also report on the characteristics of their water in terms of taste, color, and smell.

Outcomes Each data submission helps build a national inventory of pipe materials and better the infrastructure capacity of the United States. Researchers are helping identify areas at highest risk of lead in tap water to direct efforts for pipe replacements. Volunteers are raising awareness and information needed to advocate for safe drinking water.

Why we like this Every person has a right to safe drinking water. Crowd the Tap is a project from the lab of Caren Cooper, one of the authors of this book.

Project FeederWatch

In this winter-long (November through early April) survey of birds that visit feeders in backyards and other locales across North America, people of all skill levels and backgrounds periodically count the birds at their feeders. Note that while some "FeederWatchers" may see a vast variety or amount of birds, you may find yourself seeing

the same common birds every week—or perhaps you may barely see any birds at all. If this is the case, keep watching and participating and remember how important these everyday observations are for monitoring bird populations.

The annual participation fee for the project is $18. Participants receive a research kit (mailed in the fall or within three weeks of signing up during the season) that includes project instructions, a bird identification poster, a resource guide to bird feeding, a calendar, an ID number required for participation, and more.

Location North America

Materials Bird feeder, bird food, and research kit

Website Search "Project FeederWatch" on SciStarter or visit feederwatch.org

Goal To monitor backyard feeder birds

Task Register on the project website and review the handbook and instructions in the new participant research kit. Set up at least one bird feeder in an easy-to-view spot and keep it stocked with food. Keep a tally of bird species that visit your feeder. You'll start by choosing your "count days"—two consecutive days as often as once a week (but less often is fine) with at least five "non-counting" days between each two-day count. On your count days, watch your feeders as much or a little as you want and record the maximum number of each species visible at any one time during your two-day count. (See the project resources for more information on what types of birds to count and not count.) Keep a single tally across both days and submit your counts on the project website.

Outcomes FeederWatch data helps scientists track broadscale movements of winter bird populations and long-term trends in bird distribution and abundance. Results are regularly published in scientific journals and are shared with ornithologists and bird lovers nationwide. The counts you submit will make sure that your birds (or lack of birds) are represented in papers and on the project website.

Why we like this What better way to spend winter than watching wild birds outside your window while you sit in the warmth of your home?

The Great Pumpkin Project

Many of our most delicious fruits and vegetables—squash, pumpkin, zucchini, gourds, cucumber, and melons—are important crops in the cucurbit or gourd family (Cucurbitaceae). Although these crops are widely planted and economically important throughout the world, we know little about their associated microbial and insect communities. That includes both helpful beneficial insects that visit the big, lovely, sweet-smelling flowers (and in doing so, carry pollen from male to female flowers) and the harmful ones, specifically beetles, which spread a bacterial pathogen that damages these plants. Add your observations to help the project document the insects and microbes that visit different cucurbits around the world.

Location Global

Materials Cucurbit plants to observe

Website Search "The Great Pumpkin Project" on SciStarter or iNaturalist

Goal To understand the diversity of insects and microbes on cucurbit plants

Task Take photos of pumpkins and their insects and microbes and upload them to the project page on iNaturalist along with other observation details. If you save your cucurbit seeds or have an unusual variety, submit those photos as well.

Outcomes So far participants from around the world have made 7,500 observations of 289 species.

Why we like this This project allows gardeners to dig a little deeper—into the many mysteries of pests and pollinators that determine the fate of our crops.

The Great Sunflower Project

The Great Sunflower Project uses data collected on Lemon Queen sunflowers to examine the effects of pesticides on pollinators. Some bee populations have experienced severe declines that may affect food production. However, no one has ever measured how much pollination is happening over a region, much less one as big as the United States, so the information about how a decline in the bee population can influence gardens is quite limited. If you like this type of citizen science, be sure to check out the other two initiatives on the project website.

Location United States

Materials Lemon Queen sunflower seeds

Website Search "The Great Sunflower Project" on SciStarter or visit greatsunflower.org

Goal To identify where pollinators are declining and improve habitat

Task Plant organic Lemon Queen sunflower seeds (if the seeds are not organic, be sure they weren't treated with neonicotinoid pesticides). Once the flowers bloom, observe them for five or more minutes, record all pollinators that visit, and submit the data to the project website. Repeat the pollinator count at least three times while your flowers are in bloom.

Outcomes More than 100,000 members of the Great Sunflower Project have contributed important data. Each year, the number of pollinators that visit a Lemon Queen sunflower per hour is calculated. Data from 2010 to 2015 showed a steady increase in pollinator visits; by contrast, data available for 2016 showed a sharp decline, but the data and trends are not yet conclusive.

Why we like this Plant, watch, enter, repeat. That's it. And who doesn't like sunflowers?

What's in Your Backyard?

The soil found in your garden or lawn could harbor a lifesaving bio-medical discovery. Fungi found in soil create new compounds, called natural products, that may be able to treat human diseases by stopping cancer cell growth or infectious pathogens. Do you want to know what's in your backyard? The Citizen Science Soil Collection Program from the University of Oklahoma will help you find out.

Location United States

Materials Soil collection kit requested through the project

Website Search "Soil Collection" on SciStarter or visit whatsinyourbackyard.org

Goal To assist with the discovery of life-saving antibiotics

Task To contribute your data, you'll first need to request a free sampling kit from the project, which will be sent to your address within a few weeks. Once you receive your kit, find an area in your backyard that has received minimal disturbance from gardening or human activity. Remove any surface debris and collect the soil sample from the top one to three inches of your chosen spot. Seal your bag, fill out the form, and mail it back to the scientists. Once your sample is analyzed, you can check the project's interactive map to find out what fungi were hiding in your soil and learn if any of your neighbors have participated too.

Outcomes Soil samples have been sent in around the United States. In 2015, the program joined forces with the National Cancer Institute to support the scientific research of the project. A sample from Alaska uncovered a new species of a fungi that researchers are particularly interested in because it produces a compound that can block some types of cancer cell growth.

Why we like this Collecting dirt is a pretty easy assignment, but you'll be doing good with this project. Your results will be available to review and compare with others online, which is a great way to get a feel for how you've contributed.

A "Jackie and Goliath" Story

In 2008, Jackie James-Creedon and her neighbors near Buffalo, New York, were concerned after smelling something bad in the air. They figured out how to collect air samples and shared them with a scientist who analyzed the samples and found dangerously high levels of benzene, a harmful chemical that can lead to loss of bone marrow and decreased red blood cells. The New York state regulatory agency investigated and discovered that nearby Tonawanda Coke—a plant that turned coal into fuel—lied about the level of benzene it was producing. Like other industries, the plant was self-regulated. While Tonawanda Coke was reporting emissions of three tons of benzene per year, in reality it was emitting ninety-one tons of benzene per year. Criminal charges were eventually filed and the plant was found guilty of violating air pollution laws and fined more than $12 million. The judge ordered that some of the funds be used to support Jackie's mission to help others turn their data into action.

In response, Jackie founded the Citizen Science Community Resource Center, which creates soil-sampling toolkits and supports citizen scientists through the process of collecting and analyzing samples to using the data to affect real change. The kit includes instructions and everything you need (except distilled water) to properly collect soil samples: a stainless-steel hand trowel, spoon and mixing bowl, lab cleaner and brush, cooler with amber jars, flags, clipboard with paperwork to properly submit your samples for independent analysis. All of this is neatly housed in a five-gallon bucket, which you can use to wash your equipment out in the field or in your backyard. Note that the toolkit costs $250 and labs charge $50 to $200 to process samples, depending on what chemicals you want to test for. The center can also help with interpreting results and, most importantly, train and empower you to become environmental advocates using citizen science.

In more recent news, the Tonawanda Coke plant was ordered to close after Jackie's group organized around taking photos of the plant's black sooty smoke polluting their community. These photos caught the attention of regulators and the federal government once again. The company was hauled back into federal court for violating their probation, which ultimately led to the company finally shutting their doors in October 2018. This date was exactly ten years after this small group of concerned neighbors started their citizen science activism campaign against this company's egregious environmental acts. Their community is now breathing a sigh of relief that they will no longer have to endure Tonawanda Coke's toxic emissions.

Here are Jackie's tips for turning data into action:

1. Believe you can make a difference: It can start with one person or many.

2. Be inquisitive, ask questions, be relentless, never give up.

3. Form relationships with scientists and others (e.g., elected officials or media) that can help. The Union of Concerned Scientists "Community-Scientist Partnerships" is a useful blueprint for successful collaborations.

4. Talk and listen to your neighbors' stories while also doing the research before you collect any data.

5. Learn from others what worked and what didn't.

NEAR WATER

Ah, the refreshing sound of ocean waves or bubbling brooks. No matter where you live, a body of water in some form is bound to be nearby—or coming down from the sky. Citizen scientists have been monitoring water for decades as part of water quality monitoring networks. From lakes, streams, and rivers around the county, volunteers are collecting important data on the health of our waterways. If you are looking for additional resources, the National Water Quality Monitoring Council website is a great place to find a local group to make a difference in your community. Here are some tried-and-true projects that can help you get your feet wet in citizen science.

Blue Water Task Force

Nothing ruins a day at the beach quite like polluted water. We expect our waters to be clean and healthy, but heavy rains and stormwater run-off often bring pollution right to our favorite spots. You can find a local chapter of the Surfrider Foundation's Blue Water Task Force (BWTF) and collect water samples to measure for enterococcus bacteria. This type of bacteria indicates how much fecal pollution is in the water, which can cause rashes, infections, and stomach flus. The collected water samples and data you collect help fill gaps in water quality data collected by local and state officials. The information can help inform the safety of the water for swimmers and surfers.

Location More than thirty sites along the coasts of North America

Materials Whirl-Pac bags for samples, data sheets, thermometer

Website Search for "Blue Water Task Force" on SciStarter or visit surfrider.org/blue-water-task-force

Goal To monitor water quality at local beaches, bays, and ponds

Task Find your closest BWTF site and contact the local coordinator via the project website for instructions on how to collect and submit water samples to ensure coasts are safe for recreation.

Outcomes Each BWTF chapter has used their data in different ways. One chapter in Washington was able to remove a long-standing swim advisory sign by showing that the water quality in the area had improved. A chapter in Florida used their data to engage the community in action for their water. The Marin County chapter called attention to the aging sewer infrastructure, which was leading to leaks and spills during rain events, and eventually new inspection ordinance and incentives programs were developed to address the infrastructure issues.

Why we like this The BWTF is a great example of a national project having local implications. Your data can inform national trends about water quality, but also serve as important data in local water bodies. The outcomes of this project are scientific, civic, educational, and more.

CoCoRaHS

Want to participate in citizen science every day? Join CoCoRaHS (Community Collaborative Rain, Hail, and Snow Network) and measure the rainfall every day at your house, school, or place of business. CoCoRaHS was started in 1997 after a devastating flood in Fort Collins, Colorado. During the storm, some areas of the city received more than a foot of rain while other saw much less. The localized differences in precipitation weren't visible in normal weather maps and resulted in delayed warnings. This tragic event showed the need for more localized, on-the-ground weather reporting. CoCoRaHS's network of observers across North America fills this gap by contributing crucial precipitation records to forecasters and more.

Location The United States, Canada, and the Bahamas

Materials CoCoRaHS-approved high-capacity four-inch-diameter rain gauge (see project website for recommendations)

Website Search "CoCoRaHS" on SciStarter or visit cocorahs.org

Goal To provide weather data to meteorologists

Task Create an account on the project website and complete an online or in-person training about protocols, which will include tips on where to correctly site your gauge and when to collect the data. Record your daily precipitation records and remember that zero is an important data point so be sure to record that too. In addition to daily precipitation records, you can also mark down significant weather events in your community or measure snow and hail.

Outcomes CoCoRaHS is one of the longest-running project featured in this book. Over the last twenty-plus years their network has grown to more than 20,000 active observers. Forecasters, water managers, researchers, and hydrologists use data from CoCoRaHS observers every day, and CoCoRaHS data is used in scientific publications as well.

Why we like this The CoCoRaHS network is big—you'll be surprised to learn how many of your neighbors are also measuring their rain. It's a great way to contribute data that makes a difference in your community. CoCoRaHS data is commonly used to inform the National Weather Service of when to make severe weather warnings.

Marine Debris Tracker

Long walks on the beach can be quickly ruined by the sight of trash, which can come from pollution along the coast or inland. The Marine Debris Tracker app is a tool for organized groups of citizens or individuals to contribute to a global database of where, when, and what type of litter and marine debris are found. You'll log specific items of trash—like glass bottles, cigarettes, plastic bottles, and plastic bags—including the number and location of items, then safely pick up and remove the litter to keep your beach healthy. You can view all data plotted on an online map and search for sightings of specific items, like all plastic bottles. The data, part of the NOAA Marine Debris Program, is shared with scientists across the country.

Note: No data connection? No problem. The app will work offline and you can submit data when you regain service. This facilitates collection in remote areas. You can queue your submission so you make sure you don't lose data. Before going remote with the app, open it to sync it with the database and lists before going into the field (then close and reopen it when you plan to use it in the field).

Location Global

Website Search "Marine Debris Tracker" on SciStarter or visit marinedebris.engr.uga.edu

Goal To improve the health of water and beaches around the world

Task Visit the project website and download the Marine Debris Tracker app. To map the trash you find on coastlines and in waterways, you will be asked to choose a list of litter items in the app before you start tracking. (If you are working with a program coordinator, there will be a specific list for you to choose; if you are an individual user, simply use the top NOAA list as the default.) When you see a particular item, find it on the list and press the green log button to capture the item name, date/time, and GPS coordinate.

You may also add a description of the items or take a photo. Once you have logged all the items you have found (and hopefully picked up and recycled or disposed of properly), hit "view and submit." On this screen, you can see the data you have collected and delete any erroneous items.

Outcomes Marine Debris Tracker was the first app developed for collecting data on litter and marine debris and now the database contains more than 1.2 million items of litter and debris. By facilitating the collection of data at a scale, speed, and efficiency that wasn't previously possible, we can determine potential sources of the most common litter items, predict where litter will accumulate, and come up with solutions to reduce the leakage of this litter and marine debris into the environment.

Why we like this Marine debris and plastic pollution is a growing worldwide epidemic, but sometimes it is hard to really grasp the issue on land. Mapping the trash found along our waterways not only helps scientists track pollution, but also highlights the gravity of this issue.

Secchi Dip-In

The Secchi Dip-In is an annual July event to measure water quality at lakes. You only need a simple tool, the Secchi disk (composed of alternating black and white quadrants), to get started. You will lower your Secchi disk into the lake water until you can no longer see it. This depth of disappearance is a measure of the water's transparency, which can decrease as factors such as algae growth or sedimentation increase. Transparency is a basic, reliable measurement of the impact of human activity on a lake. Recording trends in transparency from year to year may serve as an early warning that activities on land are affecting a lake's water quality.

Location Global

Materials Secchi disk (homemade or purchased)

Website Search "Secchi Dip-In" on SciStarter or visit secchidipin.org

Goal To get the public involved in collecting water quality data

Task See the project website for information on where to purchase a Secchi disk or how to make your own. Register as an individual or with a lake-monitoring program and watch the online training videos to learn how to take a Secchi dip-in measurement. Lower the disk into the water until you can no longer see it and record the depth of disappearance. Submit your data online to see how the water quality near you compares to water quality across the country.

Outcomes Results from across the United States have shown how factors like urbanization and agribusiness can affect water quality. The data is publicly available online for more researchers to discover.

Why we like this The simple but powerful Secchi disk proves that the tools scientists use are not always fancy and complex. The data collected with this project becomes part of an important long-term dataset about water quality.

Shark Sightings Database

The Shark Sightings Database from the U.K.'s Shark Trust invites people from all over the world to share their shark, skate, and ray sightings. In doing so, scientists learn

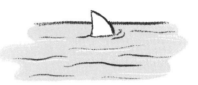

more about the colors, size, and population distributions. After uploading your observations, you can explore the website to learn more about the steps you can take to help protect these animals.

Location Global

Website Search "Shark Trust" on SciStarter or visit sharktrust.org/sightings-database

Goal To help scientists and conservationists study and protect sharks, skates, and rays

Task Share your sightings of sharks, skates, and rays on the project's online database. Include key information such as location, photos, and other observations.

Outcomes Scientists and the public are using the database to design conservation plans, inform decision-making, guide research, and educate the public.

Why we like this Sharks are vitally important to our ecosystem. But too many are hunted for their fins, which has resulted in a shark population decline causing ecological and economic ripple effects. Fewer sharks, for example, means an increase in skates and rays and other species that sharks usually prey upon. More skates and rays means fewer scallops and clams for us and fishermen. Even fish body shapes have shown measurable changes as a result of declining numbers of sharks: some fish species that are no longer threatened by sharks are evolving to have eyes and tails almost half their former size. Still, scientists need more data to fully understand the consequences of global declining populations and this project can help.

Stream Selfie

Most of our drinking water travels through streams. Better monitoring efforts are needed to understand the location and quality of water running through these streams. The Izaak Walton League of America created Stream Selfie to invite everyone to help create the first-of-its-kind national map of streams. Simply upload a picture and location of a local stream and answer a few important questions about the access to and quality of the stream. This will help determine if the stream needs targeted monitoring by volunteers and community organizations. Note: Only approach streams with safe passages and do not trespass on private property. There's no need to step in the stream.

Location United States

Website Search "Stream Selfie" on SciStarter or visit streamselfie.org

Goal To help map and monitor all the streams in the United States

Task Take a picture of the stream with or without yourself in the picture (that's what makes this a stream selfie, after all!). Next, upload your picture and share your observations directly through the project page on SciStarter. If you don't have a smartphone or you don't have service while at the stream, just note the date and time of your stream visit and upload the photo and data later.

Outcomes This new national inventory of streams was largely populated by the Girl Scouts of the USA organization, which selected Stream Selfie for their Think Like a Citizen Scientist Journey. Several troops took the project one step further and learned how to monitor the streams.

Why we like this This project makes it easy for anyone near a stream to snap and upload a picture. This alone is useful in creating a national map of streams. What's more, the organizers offer free training and materials for any citizen scientist who wants to monitor the stream's water quality.

Wet-Dry Mapping

Arizona Water Watch-ADEQ needs your help collecting locations of wet flowing or pooled segments of streams in Arizona. Water is one of Arizona's most precious resources and to protect it for years to come, officials need to know where it is and learn more about its health. Phoenix-area library patrons can check out citizen science kits for Wet-Dry Mapping that contain a GPS unit to determine location and a thermometer to collect temperature data.

Location Arizona

Materials Thermometer, GPS unit, data sheet, camera

Website Search "Wet-Dry Mapping" on SciStarter

Goal To map wet segments of streams and collect temperature data

Task Print out the project's data sheet and take it with you to a stream that is at least thirty feet long. Turn on the GPS unit and select the "satellite" icon to collect location information (latitude and longitude) from one end of the stream. Record the information on the data sheet. Walk to the other end of the stream and follow the same process. Then turn on the thermometer, confirm that the units are set to "C," and place the thermometer in the stream for at least thirty seconds. Record the temperature data on your sheet. Finish by taking pictures of both ends of the stream. Submit your information online directly through the SciStarter project page.

Outcomes Many of Arizona's citizen science water monitoring volunteers have helped scientists find pollution sources.

Why we like this Water is liquid gold in a desert state like Arizona. This project enables residents to play an active role in discovering and monitoring nearby sources of water. We also like that the accompanying kits are available through some public libraries.

AT NIGHT

_ . _ •_ ● _ . _ . _

Citizen science doesn't have to stop just because it's dark out. In fact, plenty of scientists need you to participate after hours observing nighttime critters or even the stars. These projects are great for families who see each other in the evening, camping trips with friends, or scout troops during overnight trips.

Globe at Night

Nothing beats a beautifully clear, starry night. But in many places around the world, light pollution is hiding the twinkling stars from our view. Light pollution comes from artificial outdoor lights like street lamps, stadium lights, and porch lights. Nearly three-quarters of city dwellers have never seen a pristine dark sky and many don't know what they're missing. But light pollution doesn't only affect humans—it can also do widespread ecological damage. Too much light can disrupt nocturnal animals, confuse migratory animals, and have harmful physiological effects on the rhythm of life.

This project seeks to raise awareness about increasing light pollution and quantify it by having citizen scientists measure the darkness of the sky near them. All you need is a smartphone to start making observations. You will look up at the night sky and mark its darkness based on how many stars are visible. You can also use a Sky Quality Meter (find more information on the project website) to record additional data.

Location Global

Website Search "Globe at Night" on SciStarter or visit globeatnight.org

Goal To raise awareness about light pollution around the world

Task Use the project website to find the date and constellation that the Globe at Night scientists are asking you to observe as well as the latitude and longitude of your location. Go outside more than an hour after sunset (8 to 10 P.M.) but before the moon is up. Let your eyes become used to the dark for ten minutes before your first observation. Match your observation to one of seven magnitude charts available online and note the amount of cloud cover. Report the date, time, location (latitude/longitude), the chart you chose, and the amount of cloud cover at the time of observation. Make more observations from other locations, if possible.

Outcomes In 2017, participants made more than 15,000 observations from 106 countries and all 50 states. Most of these observations showed some light pollution and less than 300 of the observations had fully starry skies. You can use the project's online interactive map to discover light pollution in your neighborhood and compare your observations to thousands of others around the world.

Why we like this Globe at Night has well-defined and straightforward protocols that are equally great for participating in at home or as a group enrichment activity during an evening event.

ZomBee Watch

The parasitic zombie fly (*Apocephalus borealis*) is laying its eggs in honeybees. This infection causes the "zombees" to behave like moths, leaving their hives at night on a flight of the living dead in search of lights, where they get stranded and die. The ZomBee Watch project scientists at San Francisco State University need your help to safely record and report where the zombie fly is infecting and killing bees. This project might take one day, if you don't find any bees, or up to thirty days if you do find a bee. That's typically how long it will take for a pesky zombie fly to emerge from its poor honeybee host. Note: This project is not recommended for anyone who has a bee or wasp allergy without careful consideration.

Location North America

Materials Light trap (optional)

Website Search "ZomBee Watch" on SciStarter or visit zombeewatch.org

Goal To help discover where zombie flies are infecting bees

Task Look for honeybees under your porch light in the morning, under street lights, or stranded on sidewalks. (Or, if you are a beekeeper, the most effective way to detect zombees is to set up a light trap near a hive—see the project website for details on making a simple, inexpensive light trap from hardware store materials.) To test for the presence of zombie fly infection, put any honeybees you collect in a container and observe them periodically. Infected honeybees will die, and in about a week brown pill-like fly pupae will emerge from the deceased bee; the adult flies will appear a few weeks later. Take pictures of the bees you collect. Submit your photos and observations directly through the SciStarter project page. Remember that not collecting any bees is also important data to report.

Outcomes Thanks to this project, scientists have confirmed the presence of the zombie fly parasite in many states and created the first-of-its-kind map of infected bees.

Why we like this If a zombie fly infection became widespread in a particular region or state, many bees would die and their hives would disappear. Because bees, which pollinate more than 80 percent of flowering plants, are already declining in population, the focus of this project is timely and important.

OUT AND ABOUT
IN YOUR COMMUNITY

Next time you are spending time outside, get ready to document nature as part of a citizen science project. If you're on a hike, slow down and enjoy the view—you never know what you might see that will help scientists. From counting birds in an urban green space to setting camera traps in a forest, the projects in this section run the gamut of where you can find nature and how much time you can commit to citizen science. No matter what, you will get outside and immersed in the world that surrounds us.

Budburst

What's the story of plants? With the Chicago Botanic Garden's Budburst project, you can help uncover and tell this story by observing plants near you. Start by familiarizing yourself with the signs, or phenophases, of five different plant groups: deciduous trees and shrubs, conifers, evergreen trees and shrubs, grasses, and wildflowers and herbs. The Budburst website offers resources to help you learn about each type and the signs you're looking for, such as flowering, leafing, fruiting, and autumn leaf color and drop. These important milestones tell scientists how plants respond to changes in their environment.

Location United States

Website Search "Budburst" on SciStarter or visit budburst.org

Goal To help scientists by observing seasonal changes in plants

Task Create an account on the project website and review the available online materials. Select which plant you are going to observe and whether you will be able to observe this plant for its entire life cycle or if this is a one-time observation (both have value). Look for its changes, like first flower, changing of leaf color, fruiting, and more. Log in to submit your observation reports.

Outcomes More than 10,000 citizen scientists have contributed their observations to Budburst and this data has been used in scientific publications to inform research about phenology.

Why we like this Budburst has easy-to-follow protocols along with ample resources for educators that connect the project to the classroom. Budburst shows science in action for students of all ages.

Bumble Bee Watch

Did you know more than 250 species of bumble bees exist around the world? And there's a lot more to that buzzing bee than you might think. With Bumble Bee Watch, you can participate in the collaborative effort to track and conserve bumble bees across North America. This project is presented in partnership with the Xerces Society, Wildlife Preservation Canada, University of Ottawa, Montreal Insectarium, York University, BeeSpotter, and the Natural History Museum in London.

Location North America

Website Search "Bumble Bee Watch" on SciStarter or visit bumblebeewatch.org

Goal To help track North America's bumble bees

Task Create an account on the project website and take some time to review the tips and resources to learn about the species that live near you. When you spot a bumble bee, snap a photo and upload your observation to the website. Be sure to include any additional notes about what species you saw and where you saw it. Your species identification will also be verified by an expert.

Outcomes In 2017, the rusty patched bumble bee was added to the endangered species list—the first bee to be listed as endangered since 1973. The assessment report that ultimately led to the endangered listing for the rusty patched bumble bee was developed from high-quality data that came, in part, from this project. You can continue to help bumble bees by contributing to Bumble Bee Watch. Maybe your data will be used to list the next species as endangered?

Why we like this Bumble Bee Watch is a simple project with powerful results as the data collected continues to be used to inform science and policy. We also like this project because, well, we like bumble bees.

Celebrate Urban Birds

Many people don't think that nature and cities can coexist. But again and again, we find examples of biodiversity thriving in urban areas. Studying urban birds, like the ubiquitous pigeon, is one way to examine how nature thrives in our cities. With this project from the Cornell Lab of Ornithology you can discover what lives in your neighborhood. It's okay if you don't know many bird species—you will only need to focus on sixteen common birds and the project website has helpful identification resources.

Location North America

Website Search "Celebrate Urban Birds" on SciStarter or visit celebrateurbanbirds.org

Goal To help ornithologists learn about urban birds

Task Visit the project website to view lists of birds in your region and tips on identification. Next, pick a time and place to observe birds. You will need to observe the birds at the same time and in the same place, so make sure you can easily come back to this spot. Your observation area will need to be roughly fifty by fifty feet (about half a basketball court). Watch birds for ten minutes and identify what birds you see in your spot. If you don't see anything, that's okay—zero is important too. Repeat your observations two more times for a total of three observations, preferably within the same week. Submit your data online.

Outcomes Observers across North America have submitted more than 30,000 checklists of

eBird

This real-time online checklist program has revolutionized the way the birding community reports and accesses information about birds. Launched in 2002 by the Cornell Lab of Ornithology and National Audubon Society, eBird provides rich data sources for basic information on bird abundance and distribution in different areas and in different times. The observations of each participant join an international network of eBird users. The project then shares these observations with a global community of educators, land managers, ornithologists, and conservation biologists. This data will eventually become the foundation for a better understanding of bird distribution across the Western Hemisphere and beyond.

Location Global

Materials Binoculars and field guide

Website Search "eBird" on SciStarter or visit ebird.org

Goal To collect, archive, and search bird observations

Task Create an account on the project website and download the eBird Mobile app, if desired. Go birding and follow the prompts on the project website or app to enter information about where, when, and how you went birding. Fill out the checklist to record what birds you saw.

Outcomes eBird data has contributed to land and wetland management and conservation plans and more than 100 scientific publications. It has also helped birders find birding hotspots.

Why we like this This project produces the largest datasets for mapping the distribution of all wild bird species. It can be done anywhere, anytime, and by anyone who can identify wild bird species.

eMammal

The goal of this project is to collect standardized data about the animals that are often unseen to humans. Camera traps are placed in the wild and automatically take a photo when something crosses its path.

Current active projects are in North Carolina, New York, and Northern Virginia. It's important to recognize that setting a camera trap with eMammal is different than a casual or recreational camera trap. The project scientists might ask you to set the trap in a random location and away from any set baits. You'll also want to set the camera trap at knee height to capture big and small animals. If no active project exists near you, you can still contribute camera trap photos to their independent, global project. Another way to participate if you can't set up a camera trap is to help analyze photos from the project online through the Zooniverse platform.

Location North Carolina, New York, and Northern Virginia

Materials Camera trap

Website Search "eMammal" on SciStarter or visit emammal.si.edu

Goal To collect, store, and share camera trap data with citizen science

Task Find a project that you'd like to contribute data to and connect with the project coordinator to make sure they need volunteers. Complete an online training module to ensure your data is collected using standardized methods. The local coordinator will then contact you with next steps for setting up your camera trap.

Outcomes Most of the data is available online for download and analysis. (Note that some projects have data embargos and photos of endangered species are not made available to prevent poaching.) Collected data has helped researchers track wildlife and estimate population sizes.

Why we like this While many citizen science projects are entry level and do not require much equipment, eMammal is a great project for those that want to participate in a longer-term project. Plus, it's a perfect fit for those that hike or hunt in the woods regularly.

GLOBE Observer

Overhead, hundreds of satellites are collecting data about Earth. But what does that data look like to humans on the ground? How can we tell that satellites are seeing what we see? With NASA's GLOBE Observer program, you can help ground-truth satellite data by making on-the-ground observations. There are several main data collection protocols as part of the GLOBE Observer app: Clouds, Land Cover, and Mosquito Habitat Mapper. Find them all at SciStarter.org/NASA.

Clouds

Have you ever sat outside and experienced the instant cooling effects of a cloud covering the sun? Imagine that feeling on a global scale with the clouds all around the world. Clouds are important regulators of Earth's temperature. Satellites orbiting Earth collect data on cloud cover but cannot distinguish small and local changes in cloud cover like humans can. Your observations will help ground-truth the images and data taken by satellites. Don't worry if you don't know all the cloud types—the app will provide helpful guides to aid your identification.

Location Global

Website Search "GLOBE Observer: Clouds" on SciStarter or visit observer.globe.gov

Goal To help scientists understand the sky from above and below

Task Download the GLOBE Observer app on your smartphone or tablet. The app will lead you through the process of data collection. You will start by making simple observations about the sky (like color, clarity, and types of clouds) and your general surroundings (like whether the ground is wet or if there are leaves on the trees). Then, you will take guided photos of the sky in all four directions plus up and down. Observations can be made at any time of day, but it is particularly valuable to make your observation when NASA satellites are overhead. The app will tell you when that is and you can even set

reminders. After submitting data through the app, you will be able to see how your observation compares to the satellite view.

Outcomes As of 2019, 24,000 people had submitted 215,000 clouds observations through GLOBE Observer. An initial publication compared volunteer-collected data to satellite data to identify the strengths and shortcomings of passive cloud remote sensing from space. Ongoing research is examining the evolution of clouds during the 2017 North American solar eclipse and incidence of cloud cover seen in citizen science data compared to satellite data.

Why we like this This project takes something as simple as observations of clouds and turns it into scientific data. It is also accessible to anyone around the world because we all see the sky at some point during the day.

Land Cover

Photographing the land around you may seem simple, but most satellite images cannot get the precise details of the land cover in your community. These details are crucial, especially in areas where high-resolution maps do not exist. Land cover transformations due to climate change or shifts in land use can affect rainfall and temperature, which can ultimately influence a community's risk for certain natural disasters like flood, drought, and landslides. The tool was designed to collect a series of photographs tagged with location and basic land cover type to be used in verifying satellite-based land cover maps.

Location Global

Website Search "GLOBE Observer: Land Cover" on SciStarter or visit observer.globe.gov

Goal To get a better picture of land cover on Earth

Task Download the GLOBE Observer app on your smartphone or tablet. The app will lead you through the process of data collection. First you will find a location with about fifty meters of clearance in each direction from where you will stand. Photograph and answer additional questions about the land cover that you see, like the types of trees and their prominence. That information will be used to calculate the overall land cover for the area observed. Submit your data and compare it to existing satellite data.

Outcomes Scientists are using this data to improve global land cover maps. Since launching in September 2018, more than 400 people have submitted nearly 2,400 land cover observations.

Why we like this Even though satellites are incredibly useful, certain data is still best collected and observed by humans. The simple project will open your eyes to the types of land surrounding you every day.

Mosquito Habitat Mapper

Despite their tiny size, mosquitoes kill more than 2.5 million people per year by spreading diseases like malaria. Satellites circling Earth can help identify the environmental variables that predict where mosquitoes might be, but on-the-ground observers must confirm their presence. Mosquitoes can breed in places where water collects or pools. Their breeding habitats therefore rely on precipitation and land cover. These factors are measured in the first two GLOBE protocols, so this final protocol helps collect crucial data on the precise areas that mosquitos may be breeding. Citizen scientists who participate in this project are actively reducing the risk of mosquito borne disease in their communities.

Location Global

Materials For sampling and photographing mosquito larvae you will need a cup, mosquito dipper, or bulb syringe; white plastic or paper plate; and a macro lens for your phone or camera with a minimum 35x magnification

Website Search "GLOBE Observer: Mosquito Habitat Mapper" on SciStarter or visit observer.globe.gov

Goal To reduce the risk of vector-borne disease in your community

Task Download the GLOBE Observer app on your smartphone or tablet. The app will lead you through the process of data collection. In this protocol you will locate and photograph potential mosquito breeding sites. This could include natural or artificial water sources that may contain mosquito larvae, like water pooling in a bucket or an old tire. If desired, you can take a small sample of water, photograph the mosquito larvae on a white surface, and count and identify the larvae type. Lastly, you will be prompted to eliminate the habitat you found by dumping out the water or covering it with a lid or net.

Outcomes As of March 2019, 2,800 citizen scientists from 69 countries have submitted nearly 9,500 observations from nearly 3,500 sites. The Mosquito Habitat Mapper community is partnering with scientists who are using mosquito data to develop mathematical models to predict outbreaks of West Nile virus in the United States. Other scientists in Hawaii are employing citizen science data to protect endangered bird species from outbreaks of avian malaria. Citizen scientists have also used the data they collected using this app to reduce risk of disease in their own community. For example, students from a school in Peru identified breeding sites of mosquitoes that were not known by the municipality. By communicating their results with the mayor's office, their data was used to target and eliminate breeding grounds by the municipal mosquito control program. As the data set from GLOBE Mosquito Habitat Mapper increases its spatial and temporal reach, NASA scientists will be better able to understand how mosquito populations respond to changes in weather, vegetation, and climate.

Why we like this Mosquitoes are more than just a summertime nuisance. They carry deadly diseases and with changing climate and rising temperatures, we are seeing mosquitoes in new places. In turn, mosquito-borne diseases are spreading. Mosquito Habitat Mapper combines making observations with the actionable opportunity to decommission potential breeding sites.

iNaturalist

Keep an online nature journal by taking photos of anything wild and help collect important biodiversity data at the same time. Your biodiversity observations will become important data for scientists. Maybe you found a squirrel while it was foraging or a plant that was covered in an unusual fungus. If you want to make biodiversity observations with your friends, family, or community, consider organizing a bioblitz—a fun way to collect as much biodiversity data as possible in a single location. The iNaturalist website hosts a number of how-to guides to get started.

Location Global

Website Search "iNaturalist" on SciStarter or visit inaturalist.org

Goal Contribute to a global database of biodiversity data

Task Create an account on the project website and download the iNaturalist app, if desired. (Tip: add your iNaturalist username to your SciStarter account settings. You'll earn credit for your contributions.) If you see something wild on your next hike (or just while exploring your own backyard or the corners of your home)—including plants, animals, fungi, or even protozoans—snap a photo and upload it to iNaturalist. Make sure to keep track of important data like time, location, and any notes to make a complete observation. Identify what you see if possible. If you don't know, the iNaturalist community will help crowdsource an identification. There's even new AI technology that will provide suggested identifications based on the appearance of your observation, its location, and what similar things have been seen nearby.

Outcomes More than 13.8 million observations of 181,00 species are currently uploaded to iNaturalist . . . and counting. Observations marked as "research grade" are shared with the Global Biodiversity Information Facility (GBIF), an open-data research infrastructure for global biodiversity data that is accessible to scientists around the

ISeeChange

What does change look like near you? Have you noticed flowers bloom-ing earlier than usual? Has extreme weather and flooding affected your commute? You can join a community of climate and weather journalists with ISeeChange. It is simple to get started. All you need to do is observe something changing near you or something that just seems unusual. If you're unsure what to look for, take a look at the featured investigations on the project website that target climate-related trends such as flood-ing, heatwaves, and bugs. Some may even be unique to your location.

ISeeChange partners with NASA's Orbiting Carbon Observatory 2 mission to ground-truth the data from satellites with observations on the ground. Your observation will be matched with data from that date and time including temperature, cloud cover, and carbon dioxide lev-els in the atmosphere, which will help scientists see and understand the impact of the changing climate on everyday life. You can also join community investigations that seek specific data in your community. From rising tides to urban heat waves, your data will help communities create climate adaptation plans.

ISeeChange was founded by Julia Kumari Drapkin, a radio, televi-sion, and multimedia producer who uses her knack for storytelling to share the local and global impacts of climate change. Her team uses the observations contributed to ISeeChange to write in-depth pieces on changes in local environments.

Location Global

Website Search "ISeeChange" on SciStarter or visit iseechange.org

Goal To connect and empower communities to investigate weather and climate change

Task Create an account on the project website and download the ISeeChange app, if desired. Post sightings and observations, including photos when possible, on how weather and climate affect your daily life. Check back often to help compare and contrast experiences and to find out what others see near you or developing trends to watch.

Outcomes The ISeeChange community, in addition to having their own localized records of change, has identified environmental trends months in advance of scientists, journalists, and government reports; ground-truthed data that engineers and infrastructure designers rely on; provided critical community insights into climate-resilience planning; and provided emergency managers with real-time updates during weather events.

Why we like this This project is based on the simple premise that everyone talks about the weather and asks questions like, "What is going on with this weather today?" Those anecdotes can provide powerful data about how the changing climate influences your everyday life.

Mushroom Observer

Mushrooms are funky fungi that are mysterious to many scientists. Although we've all seen and probably eaten a mushroom, it is estimated that less than 5 percent of the world's species of fungi are known to science. That leaves a lot to discover. One reason fungi are so elusive to scientists is that they are difficult to identify simply by sight. Most of the organism known as a fungi is underground and only the fruiting body, known as the mushroom, is seen from the surface. To make matters even more confusing, not all fungi produce mushrooms. Fungi like bread molds and toe fungus (yuck!) do not produce a mushroom state. This project helps bridge the gap between the unknown and the experiences of citizen scientists and amateur naturalists around the world.

Location Global

Website Search "Mushroom Observer" on SciStarter or visit mushroomobserver.org

Goal To expand scientific exploration of mushrooms

Task Create an account on the project website and take some time to familiarize yourself with the mushroom terminology. This will help you make better observations and take full advantage of the site. Next, go outside and find some mushrooms. You may be able to find them in your backyard, schoolyard, or park. When you find a mushroom, take a photo and upload it to Mushroom Observer. Try to identify your mushroom or ask the community for help.

Nature's Notebook

When was the last time you took a close look at nature? Sometimes we need a gentle reminder to simply stop, observe, and learn. With Nature's Notebook, from the USA National Phenology Network, you'll spend time outdoors slowing down to enjoy nature. Up-close observations of plants and animals will deepen your connection to the great outdoors, plus you'll be helping science.

Nature's Notebook is especially interested in phenology—the timing of plant and animal phenomena like sprouting leaves, the first spring bird songs, and more. Close observations of plants and animals can help scientists observe phenological changes that may be overlooked. Nature's Notebook data is often used to ground-truth the results from scientific modeling as the real-time data helps scientists see if their predictions are actually happening in the real world.

Location North America

Materials Binoculars, data sheets, clipboard, and pencil (optional)

Website Search "Nature's Notebook" on SciStarter or visit usanpn.org/natures_notebook

Goal To help scientists create a broad picture of how nature is changing

Task Create an account on the project website and download the Nature's Notebook app, if desired. Determine where you want to make your observations, identify the plants or animals you want to observe (you can use the species or campaign lists provided by the project), and observe each for about two minutes. Use the mobile app or data sheets to record your observations.

Outcomes The expansive data collected through Nature's Notebook is being used to make new discoveries and enhance our understanding of the natural world, such as showing that the typical signs of spring are occurring earlier in 76 percent of our national parks. This data has also shown that milkweed (a key host for monarchs and other pollinators) is blooming earlier, which has important implications for how pollinators use the plant.

Why we like this Nature's Notebook is a robust citizen science project with easy to follow instructions, publicly available data, educator resources, and clearly defined outcomes.

NestWatch

Whether in a shrub, tree, or nest box, bird nests are everywhere. Find one, and you can help scientists study bird biology and monitor populations of North America's birds. Every year, volunteers from across the country visit nests once or twice each week and monitor their progression from incubating eggs to fuzzy chicks to fully fledged young. Once submitted, this data is compiled and analyzed by the Cornell Lab of Ornithology's NestWatch program.

Location North America

Materials Binoculars; data sheets, clipboard, and pencil (optional)

Website Search "NestWatch" on SciStarter or visit nestwatch.org

Goal To help scientists monitor North America's nesting birds

Task Create an account on the project website and download the NestWatch app, if desired. Take some time to read, study, and sign the code of conduct to become a certified NestWatcher so you can find and monitor nests without disturbing the birds. Once you find a bird nest, visit it every three or four days and record what you see. Submit your observations in real time with the app or record your observations on the data sheets and then add the information to the website later.

Outcomes NestWatch data has been used in numerous peer-reviewed publications on topics from local changes in nesting to large-scale trends. Data from citizen scientists has helped inform how temperature and precipitation affect the first eggs of bluebirds in Ohio and other studies have shown how broad seasonal and latitudinal trends impact clutch size.

Why we like this NestWatch helps people of all ages and backgrounds connect with nature. The information that NestWatchers collect allows us to understand how various threats such as environmental change and habitat destruction impact breeding birds. Armed with this knowledge, we can take the necessary steps to help birds survive in this changing world. Data on the timing and success of breeding birds are key indicators of environmental health. It's fun, easy, and it helps the birds.

Project Squirrel

Squirrels are animals that we almost all see, but rarely think twice about. Project Squirrel encourages everyone to pay more attention to these common animals. All you need to do is count the squirrels in your neighborhood and submit your data online. You'll need to know some basic information about the squirrels including the type of squirrel, like red, fox, or gray (did you know that black squirrels are really just gray squirrels?). You may be surprised to learn the diversity of types of squirrels in your neighborhood.

Location Global

Website Search "Project Squirrel" on SciStarter or visit projectsquirrel.org

Goal To help scientists better understand tree squirrel ecology

Task Find a spot to observe squirrels, count how many squirrels you see, record the species (i.e., gray, red, or fox), and take a photo. Make additional observations including what trees are present, the location type, if the squirrels were feeding, and if any threats like hawks, coyotes, dogs, or cats are in the area. Remember, not seeing any squirrels is still important data to record. Report your findings directly through the SciStarter project page. Come back often to the same observation spot to continue tracking squirrels in your community.

Outcomes Project Squirrel has helped illuminate the distribution of different types of squirrels in urban areas. Since the project asks for additional data, like the types of trees nearby, landscape type, food presence, and more, scientists have started to understand how these additional factors correlate to where squirrels live.

Why we like this You can participate in Project Squirrel anywhere from an urban neighborhood to a forested park to your own backyard. The data you collect is simple and easy to understand.

Southwest Monarch Study

Southwest Monarch Study is dedicated to monarch butterfly research and conservation in the Southwest United States. The project provides training to help citizen scientists tag monarchs on their migration and to grow and monitor monarch habitats.

Location Arizona, New Mexico, Nevada, Utah, western Colorado and the California deserts

Materials If needed, request monarch tags through the project

Website Search "Southwest Monarch Study" on SciStarter or visit swmonarchs.org

Goal To learn about monarch butterfly migration and breeding in the Southwest

Task Visit the project website for more information on the many ways that you can participate. Options include monitoring monarch breeding habitats; reporting sightings of adult monarchs, eggs, caterpillars, and larvae; learning to tag monarchs in the late summer and fall or joining a tagging trip; looking for monarchs during the winter to help monitor their movements in the southern deserts; and helping spread the word about the need to grow more monarch habitats in the Southwest.

Outcomes More than 600 citizen scientists (and counting) have tagged thousands of monarchs in the Southeast. Recovery of tagged monarchs has shown that in Arizona not all monarchs migrate, and those that do migrate to both Mexico and California. More nuanced data and results are available on the project website.

Why we like this All tagging data is shared openly with researchers, conservationists, and the public.

The Lost Ladybug Project

Ladybugs might all seem similar but take a closer look and you'll see that a variety of spots and colors showcase the biodiversity of these insects. The species composition of ladybugs across North America has been rapidly changing as non-native species replace native species. The Lost Ladybug Project encourages citizen scientists to submit their photos of ladybugs from across North America to help understand where all the ladybugs have gone.

Location North America

Website Search "The Lost Ladybug Project" on SciStarter or visit lostladybug.org

Goal To map ladybug species and help protect and restore native species

Task Visit the project website for resources on how to find, collect, and photograph ladybugs. It can help to put the ladybugs in the freezer (for no more than five minutes) to slow them down before photographing them. To get the best photos, place the ladybug on a light gray or white background and snap your photos (the more photos the better) using the close-up setting. Upload your photos to the project website along with a brief description of where and when you found the ladybugs and how you collected them.

Outcomes As of 2019, participants had contributed more than 38,000 ladybugs to the project. The data is all available on the website, so you can explore the maps and compare the current data to historical results. The Lost Ladybug Project has also led the restoration of native species of ladybugs through reintroduction efforts.

Why we like this This project uses a seemingly ubiquitous insect to answer important questions about changing species makeups across North America. It serves as an important reminder that even though we see things like ladybugs all the time, there is still more to discover upon a closer look.

TreeSnap

Can you identify a healthy hemlock, chestnut, ash, elm, or white oak tree in a forest? If so, scientists need your help finding these resilient trees to study. Learning more about them will provide clues on why some trees are destroyed by invasive diseases while others are not. This app was developed as a collaboration between scientists at the University of Kentucky and the University of Tennessee.

Location Global

Website Search "TreeSnap" on SciStarter or visit treesnap.org

Goal To locate healthy hemlock, chestnut, ash, elm, or white oak trees for scientists to study

Task Create an account on the project website and download the TreeSnap app. Follow the app's instructions to start adding locations and identities of healthy trees you see. Remember, it's more helpful for the scientists if you can accurately identify these particular tree species.

Outcomes Scientists affiliated with organizations including the USDA, the Forest Restoration Alliance, and the UConn Plant Computational Genomics Lab have already started using the submitted data.

Why we like this TreeSnap does not serve as a tree identification resource but it provides key identification features as a reference for each species. The app takes privacy into account—yours and the tree's—by not publicly displaying exact locations of the trees in case they're on private land. However, you, the administrators, and scientists can see the precise location of the tree you added to the database.

Western Monarch Thanksgiving Count

Led by the Xerces Society for Invertebrate Conservation, this project takes place in California during the three weeks spanning Thanksgiving as well as around New Year's. Monarchs from many other states travel to these wintering sites on the California coast so they are perfect locations to observe and survey populations.

Location California coastline

Materials Binoculars, data sheets

Website Search "Western Monarch Thanksgiving Count" on SciStarter or visit westernmonarchcount.org

Goal To monitor monarch butterfly populations as they cluster along the California coast

Task Create an account on the project website and connect with a regional coordinator to receive hands-on training. Select a specific site where monarchs overwinter to monitor and complete a one-page habitat assessment so the organizers can track environmental changes at the site. Visit your site during the early morning (when monarchs are non-mobile) and use binoculars to look at clusters and estimate the number of monarchs per cluster. Record your observations and submit your data sheets to the project to tally population numbers.

Outcomes The organizers have more than twenty years of data showing that monarch populations have declined in the western states from 1.2 million monarchs reported in the late 1990s to less than 290,000 reported in 2018.

Why we like this Many pollinator-tracking projects are chance-based, meaning that if you happen to see a pollinator, you can record and share data. This project provides an important and efficient snapshot of the monarch population by inviting citizen scientists to view clusters and colonies, making it easier to find and count hundreds of thousands of monarchs year after year.

CHANCE
PARTICIPATION

Sometimes you just have to be in the right place at the right time to participate in a citizen science project. Be sure to read about these serendipitous projects so you'll be able to come back to them if the opportunity happens to arise.

California Roadkill
Observation System

Make road trips extra fun by adding citizen science into the mix. Ever seen a squashed squirrel on the side of the road? Roadkill is an unfortunate result of how humans and animals have come to coexist. Although avoiding every episode of roadkill is impossible, scientists need your help identifying incident hotspots to understand what factors contribute to roadkill and ways to prevent it. We encourage passengers to take part in this project while drivers keep their eyes on the road—it's a great way to keep backseat drivers at bay or entertain the kids.

Location California

Website Search "California Roadkill" on SciStarter or visit wildlifecrossing.net/california

Goal To understand the factors that contribute to roadkill

Task Create an account on the project website in order to submit roadkill sightings. Once you spot roadkill, try to identify exactly what it is or at least what type of animal (i.e., bird, mammal, reptile). It's okay if your identification is not certain; you can mark your level of certainty on the data collection form. Next, where did you see the roadkill? If on the highway, look for mile markers and exits to help you identify your location. Within a city, look for the closest address or cross streets. Finally, if you're able to safely snap a photo, you can do that and upload it to the site.

Outcomes This project is part of the Road Ecology Center at the University of California, Davis. Each year, they use the data from citizen science contributions to publish a report outlining roadkill hotspots to inform decision-makers and policy-planners across the state. The observations are also added to an interactive map so you can explore the type and location of roadkill. Stay tuned for potential expansion of this project around the United States and the world.

Why we like this We don't often think about roadkill, but it can seriously affect animal populations. By asking people to observe the roadkill around them, it opens up the opportunity to protect animal populations through changes in our transportation systems.

Dragonfly Swarm Project

Have you ever seen a single dragonfly flying near a pond? What about hundreds to thousands of dragonflies, sometimes of different species, swarming in one place? Some people are lucky enough to see this phenomenon. It doesn't last long and researchers don't know much about the behavior yet.

Chris Goforth, an aquatic entomologist (that means she studies bugs in and near water), was so perplexed by her first-time witnessing this behavior that she wrote about it on her blog. Suddenly, she was receiving comments from other people who had witnessed the same odd behavior and were curious about what it meant. This was the start of the Dragonfly Swarm Project. Chris created this citizen science project to systematically collect these reports for people across the country. She received 650 reports in the first summer. If you're lucky enough to witness this behavior, make sure to enjoy the moment and observe as much as you can.

Location Global

Website Search "Dragonfly Swarm Project" on SciStarter or visit thedragonflywoman.com

Goal To contribute to large-scale research of dragonfly swarm behavior

Task Spot dragonfly swarm behavior and report your sightings on the project website. Include important details such as estimates of the number of dragonflies, number of species, time and date of the swarm, and recent weather conditions.

Outcomes The majority of swarms reported each year occur during dragonfly migration. Based on the data contributed to the project thus far, disturbances (anything from approaching hurricanes to droughts to people mowing their yards to major thunderstorms) appear to play a major role in the formation of most dragonfly swarms reported outside of the migration season.

Why we like this This project is challenging to participate in since you have to be in the right place at the right time. But those that witness the elusive dragonfly swarm behavior are so excited to find this project to share their experience.

Landslide Reporter

Imagine the sound of a roaring landslide. Or how narrowly escaping one must feel. While landslides occur around the world, no global understanding exists as to when and where they will occur. Large incidents are reported and covered by the media, but small incidents may only show up in local newspapers or are perhaps only known by the few people who live nearby. That's why NASA created the Cooperative Open Online Landslide Repository (COOLR) with the opportunity for the public to participate. Note: When investigating a landslide, the most important thing is to be in a safe place—never conduct field work or look at a landslide up close unless you are an expert.

Location Global

Website Search "Landslide Reporter" on SciStarter or visit landslides.nasa.gov

Goal To build an open, global landslide data resource for science and decision-making

Task Submit landslide events experienced in person or found in a newspaper article, on the local news, or in an online database to the Landslide Reporter website or download the mobile app. Include the setting, impacts, and other details. After the team at NASA verifies and approves your landslide report it will appear publicly in the COOLR alongside other landslide inventories.

Outcomes Data on landslide events from citizen scientists can help guide awareness and action to protect against landslide hazards and enable researchers to study their future impact. Each landslide report submitted helps improve understanding of landslide hazards and scientific modeling of the risk in different areas around the world. Data will also be used to aid emergency response preparation for disasters.

Why we like it You can participate in this project even if you don't experience a landslide first-hand by submitting reports from news articles.

The Total Solar Eclipse

Some citizen science projects occur around infrequent events, such as the total solar eclipse that occurred on August 21, 2017. The eclipse's path of totality moved across the continental United States, where 12.25 million people live, including major cities like Salem, Oregon; Lincoln, Nebraska; Jefferson City, Missouri; Nashville, Tennessee; and Charleston, South Carolina.

Several scientists jumped on the opportunity to use the power of the millions of people watching the event to collect data. The GLOBE Observer program encouraged people to take temperature readings before, during, and after the eclipse in order to see how much the earth cooled while the sun was blocked. Citizen scientists also took pictures of the clouds and made atmospheric observations. At the time of writing, the NASA scientists were still analyzing the data from the 10,000-plus observers who made more than 80,000 air temperature readings and more than 20,000 cloud observations. This once-in-a-lifetime experience also provided a wealth of rare data for scientists.

Scientists with the California Academy of Scientists and iNaturalist encouraged people to observe the way animals reacted to the eclipse. The Nashville Zoo even used the opportunity to ask visitors to monitor zoo animals. Although it's likely that the zoo animals were mostly reacting to the guests rather than the change in the sun, the citizen scientists involved in this project observed birds flocking, frogs calling, crickets chirping, and much more.

Citizen scientists were fortunate to be a part of the 2017 eclipse. These opportunities only come around every so often, so don't forget the next total solar eclipse will pass through the United States on April 8, 2024.

CITIZEN
SCIENCE
in Schools, Libraries, and the Community

Citizen science is a natural fit for educators of all levels, whether you and your students engage in projects in the classroom or library, on a field trip, or in the lab. Beyond this book, be sure to connect with local parks, museums, and other organizations to find great local projects that are specific to your region, state, or community. You'll find that eager individuals—of all ages—and families across the world are interested in discovering and engaging in scientific endeavors to help address an issue through science and to learn something new about the world around them. Also check out the resources section of this book for helpful links and tips specific to educators and librarians.

IN SCHOOLS

_ . _ • _ ● _ • _ . _

From elementary school through higher education, citizen science belongs in the classroom. Citizen science projects provide authentic inquiry-driven questions for students to answer through the scientific process. Youth participating in citizen science projects are no longer just demonstrating the scientific process through labs and activities, they are engaged in genuine scientific practices. Youth-led citizen science allows students to take ownership of their data, share their results outside of the classroom, and engage in the complex social, scientific, and technological realms of science.

When selecting a citizen science project for the classroom, additional factors should be considered. First, some projects are specifically designed to accommodate learning outcomes as well as scientific outcomes, while others offer less support for classroom educators. Often, as we've mentioned earlier, this may be due to the limited support for project management and not a reflection of the project's intent to engage youth. In classroom citizen science, it is particularly important for educators to review the protocols, understand the materials needed, and determine the types of feedback that the project will provide. There are many great examples of teachers bringing citizen science into their classrooms, to their homeschool groups, and to other alternative learning settings. Here we've highlighted a few examples that take a system-wide approach to implementing citizen science in schools.

Middle schools throughout Broward County, Florida, achieve their problem-based learning goals through a customized citizen science portal on SciStarter. There, teachers explore and assign curated projects, track student progress, and reflect on experiences together. In the process, the district acquires valuable data to help support students' interests, experiences, persistence, and learning outcomes while measuring the collective impact their district has made through citizen science.

Similarly, North Carolina State University embeds campus life with citizen science through a customized portal on SciStarter. NC State is a Citizen Science Campus with the goal that every student encounters citizen science opportunities at least seven times during their undergraduate education. These opportunities include courses that assign citizen science volunteering as homework or class projects, undergraduate research credits, events with the student-led Citizen Science Club, featured projects in the Library's annual Wolfpack Challenge, observation hotspots, citizen science kits for sale in vending machines, and community service credits with dormitories and residential villages.

As part of an emerging trend, SciStarter is collaborating with projects, organizations, and institutions to explore and pilot approaches to track, assess, and accredit citizen scientists—particularly those who do not do citizen science in formal learning environments such as schools—for their experiences and valuable contributions to science. Don't be surprised to see citizen science listed as a desirable experience on LinkedIn soon.

Journey North

Every year thousands of animals migrate north
and south as the seasons change. With Journey North, you can track the migration
of your favorite species and make valuable contributions to understanding the
dynamics of migration. Your classroom
can also participate by planting a Journey
North Tulip Test Garden that will show how
your local climate affects plant growth and
contribute to a long-term database. Journey North is a citizen science project that is
designed specifically for educators and students, but anyone is welcome to contribute.
The website has numerous resources about
the various migrating species as well as tools for
the classroom including instruction activities, vocabulary
activities, reading strategies, and inquiry strategies.

Location United States

Website Search "Journey North" on SciStarter or visit
journeynorth.org

Goal To study wildlife migration and seasonal change

Task Create an account on the project website and select the
projects you'd like to participate in. Journey North's monitoring
projects range from monarch butterflies to American robins to gray
whales. Once you pick the species you want to monitor, you will
contribute sightings throughout their migration season and students
will share their field observations with classmates and online. If you
choose to plant a test garden, you will plant your tulip bulbs in the
fall and collect data when the plants emerge in the spring.

Ant Picnic

What do ants like to eat? Help scientists find out the dietary preferences of ants by making them a picnic. Researchers at North Carolina State University previously used citizen scientists to answer the question "What ants live where?" around the world, but now they want to know what those ants eat. The results from your ant picnic can provide scientists with important information about what nutrients ants forage for at different times of the year and in different locations. Ant Picnic is great for classrooms or community groups like scouts, nature clubs, play groups, and homeschoolers. The materials are readily available and inexpensive and online lesson plans (studentsdiscover.org/lesson/ant-picnic/) cover the background information about the project and provide step-by-step instructions for students to create and set the baits.

Location Global

Materials Extra-virgin olive oil, L-glutamine powder, sugar, water, salt, cotton balls, containers for mixing solutions, measuring spoons or scale, Pecan Sandies cookies, "Experiment in Progress" sign, data sheet, sandwich-size Ziploc bags, white 3x5 index cards, pencil, smartphone or digital camera

Website Search "Ant Picnic" on SciStarter

Goal To inform scientists about global food preferences of ants

Task To create a picnic for ants, you will first need to gather the bait materials. Next, you will prepare and mix the baits as outlined on the website and set them up in two outdoor spots: one green and one paved. Wait about an hour and record how many ants are on each bait in each location. Upload the data directly through the project page on SciStarter.

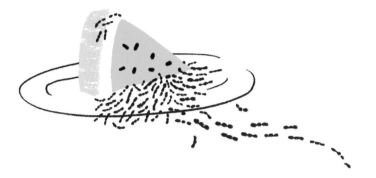

Outcomes The ant picnics results are incorporated into the largest study of global patterns in preferred resources and activity within a single group of organisms.

Why we like this Ant Picnic is a great group project and especially fun for kids. The project also shows many of the steps of setting up a scientific experiment, such as making a prediction, collecting data, and analyzing results.

Teaching with Citizen Science

For Linda, citizen science started at home when her daughter was in elementary school. They started doing citizen science together through the Lost Ladybug Project and loved how it got them outside, active, and involved in nature. After this experience, Linda's interest in citizen science blossomed in the classroom. She found a wealth of information and projects that she wanted her classroom students to be involved in. Her students began to explore their outdoor environment, finding and identifying critters. The experience was enriched even more when the scientists responded. They would set up the projector in her classroom and look at the photos, identifications, and responses from the scientists together. Sometimes the students got the right identification, other times the scientists came back with a different identification. Either way, they were discovering and learning alongside a scientist.

The classroom connection was even stronger because the data connected to the real world. Linda could tell her students that this data and their effort matters to the scientists. And they were hooked. This meant that the learning became even more meaningful. Citizen science also makes the students feel valued and important.

When it's time pick a new project for her classroom or her family, Linda starts simple. She looks at the directions and what they're asking for her to do. It has to be something accessible and easy for her to execute. Next, she has to find something that she is personally interested in and passionate about. If she finds it exciting, it's much easier to relay that excitement to her students. And finally, the equipment and tools required for the project must be affordable and scalable to her entire classroom.

Like many teachers, Linda evaluates classroom activities based on available lesson plans and curriculum standards. She uses the search filters on SciStarter to find projects that are suitable for her classroom grade level and that have teaching materials available. Often, she uses the teaching materials as a starting point for how she wants to approach the project in her classroom. The back-to-school blog series and educator's page on SciStarter has more ideas.

IN LIBRARIES

_ . ─ • ─ ● ─ • ─ . ─

Libraries and librarians are tasked with being the future-ready places for education and connection. More and more libraries are engaging in science programming from makerspaces to citizen science. Libraries provide safe, welcoming spaces to find resources to access data and information, related programming, recommended books and media, access to computers and Wi-Fi, and, in some cases, the instruments needed to do projects—such as telescopes, binoculars, air quality sensors, constellation guides, or specialized citizen science kits.

Here are a few ways librarians can introduce citizen science experiences and resources to patrons:

* Invite local subject matter experts (including citizen scientists themselves) to talk about topics related to the projects your library promotes.

* Build or leverage existing programs around the projects and events by connecting related citizen science projects as activities your patrons can engage with, either at the library or at home.

✳ Libraries are central to the success of Citizen Science Day—an annual April event designed to celebrate all things citizen science—so check out programming ideas and other ways to get involved.

✳ Most importantly, engage in citizen science projects yourself so you will have firsthand knowledge of what patrons can expect and the type of questions they may ask.

✳ Find more resources on SciStarter.org/library.

The Librarian Tasked with Engaging in STEM Programming

Ryan works at a rural library on the East Coast. He has been a longtime precipitation observer with CoCoRaHS (see page 84) after reading a newspaper article about the program while he was still in high school. Now as a librarian, he wants to bring citizen science to his patrons.

He first started with a California Academy of Sciences program called Science Action Club. This program introduced him to iNaturalist and a project for students to observe bugs in their schoolyard. As the kids ate it up and continued to want more and more, Ryan came to SciStarter to find those opportunities. Now he's doing eBird, NestWatch, and more.

When looking at programming for a library, Ryan recommends finding projects that don't require a daily task and that are relatively cheap. Integrating smartphones with opportunities to go outside have been a big success. Finally, you can sort by projects that have teaching materials on SciStarter so that you have a curriculum on which to base your project choices.

Ryan's advice for your first citizen science experience is to find a program that is easy to get started with, has simple materials, and can engage your library visitors. From there, you can build up to more advanced opportunities.

AT SCIENCE CENTERS, MUSEUMS, AND AQUARIUMS

Heading to your local science center, museum, or aquarium soon? Many institutions offer opportunities to contribute to citizen science as part of their nature walks and other public programming. Here are some examples:

* The Natural History Museum of Los Angeles County hosts a robust community science program focused on urban nature. Their museum runs several citizen science projects including GeckoWatch, RASCals, SLIME, Spider Survey, and the Los Angeles Butterfly Survey. If you visit the museum, you can participate in these projects in their Nature Garden and contribute to the L.A. Nature Map. Many of the projects are hosted on the iNaturalist platform. Also keep an eye on their evening events (community science plus cocktails!) and bioblitzes throughout the city.

* The North Carolina Museum of Natural Sciences is a great place to find and participate in a citizen science project. At the SciStarter kiosk you can learn about current opportunities and

find a new project to participate in. The museum hosts several projects including Cheese Alive, CitSciScribe, Natural North Carolina, Sparrow Swap, and more. If you'd like a guided citizen science project, visit the Prairie Ridge Ecostation, where you can find events from bird nest monitoring to leaf litter sampling.

✳ The Denver Museum of Nature and Science hosts a unique citizen science experience for visitors in their Genetics of Taste Lab. Museum scientist Dr. Nicole Garneau uses the museum lab to collect important data on taste from museum visitors. If you visit, you may be asked to sample different solutions and rate their taste or even swab your mouth for DNA. The results are published in health journals and the publications are available on the lab's website. Stop by to check out the latest project and get a taste of citizen science.

✳ The San Diego Natural History Museum hosts a permanent exhibit entitled "Extraordinary Ideas from Ordinary People: A History of Citizen Science." This exhibition looks at how naturalists through time have shaped our understanding of science. You'll get a glimpse at the museum's research collection including curated rare books, art, photographs, and more.

✳ The California Academy of Sciences offers citizen science experiences, most of which feature iNaturalist, for museum visitors and Bay Area residents. The Academy's purpose is rooted in discovering and documenting worldwide biodiversity and their citizen science projects take that focus to the local area. Join their bioblitzes to document biodiversity in tide pools, in managed watershed lands, and throughout San Francisco's urban areas. The Academy also provides resources to educators looking to implement citizen science through their Science Action Club and the Citizen Science Toolkit.

✳ The Oakland Museum of California's innovative citizen science vending machine makes getting involved in citizen science as easy as buying a candy bar. Located in the Gallery of California Natural Sciences, the vending machine contains the tools and instructions to get started in a project. The projects in the vending machine rotate based on what's featured in the exhibits. If you stop by the museum, bring $5 cash to purchase a kit.

✳ The Field Museum in Chicago offers a swath of citizen science projects that support the research of citizen scientists in Chicago and beyond. You can monitor wildlife in Illinois, measure liverwort plants, or participate in national projects like School of Ants.

LiMPETS

LiMPETS (Long-Term Monitoring Program and Experiential Training for Students) is a well-established citizen science project conducted up and down the 600-hundred-mile coastline of California. This program, open to sixth- through twelfth-grade students, is designed for schools, informal environmental education, and community groups. Participants develop a scientific understanding of the ocean by monitoring California's coastal ecosystems. You can get involved by contacting the LiMPETS coordinator in your area.

Citizen scientists help monitor sandy beach and rocky shore coastal ecosystems. At rocky shore sites, participants count intertidal organisms, including anemones, hermit crabs, sea urchins, sea stars, barnacles—all creatures that are easy to identify and also sensitive to changes in their environment. At sandy beach sites, participants count sand crabs burrowed underneath the sand. Sand crabs are important herbivores found along the entire Pacific Coast and represent a key indicator species in the sandy beach ecosystem. Many birds, fish, and other animals rely on healthy sand crab populations for their food.

Location California coastline

Materials Depends on your collection protocol

Website Search "LiMPETS" on SciStarter or visit limpets.org

Goal To provide authentic, hands-on coastal monitoring experiences for students

Task Participate in a coastal monitoring protocol, share your data, and become an ocean steward. To get started with LiMPETS, teachers and educational leaders will need to contact their local coordinator, attend a workshop, and arrange for a LiMPETS researcher to come and teach your students the protocols and identification. Once you have the materials and training, you can begin monitoring, collecting data, and analyzing your data.

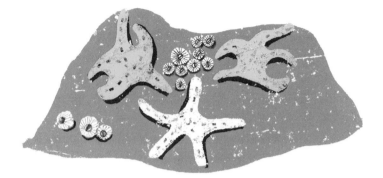

Outcomes LiMPETS data provides important insights into California's coastline ecosystems and national marine sanctuaries. The data set is valuable for assessing current trends, but also serves as a point of reference for future observations, especially in response to disasters like oil spills, habitat destruction, ocean acidification, and more.

Why we like this The LiMPETS program perfectly blends student engagement with science to create an authentic experience teaching about ecosystem-wide issues.

A Year of

CITIZEN
SCIENCE

JULY

Looking for connections between citizen science and holidays, current events, or seasons? You can plan an entire year of citizen science projects and events with this calendar. Find more at SciStarter.org/calendar.

JANUARY

_ . _ • _ ● _ • _ . _

Brr . . . is it chilly out? If it's too cold or snowy to go outside and participate in citizen science, use this opportunity to scroll through SciStarter and bookmark projects for spring. There are also lots of ways to spend time indoors with citizen science. You can contribute to Alzheimer's research from the comfort of your couch by identifying stalled blood vessels in movies of mice brains with **Stall Catchers** (page 62). Other entirely online projects include **Bat Detective** (page 52), **DeepMoji** (page 55), **Foldit** (page 57), **Galaxy Zoo** (page 58), **Smithsonian Transcription Center** (page 60), and **SETI@home** (page 59).

January 21st also marks Squirrel Appreciation Day, so go outside and find these bushy-tailed mammals in your neighborhood, park, or backyard. Contribute your observations to **Project Squirrel** (page 121) to help scientists understand the populations of squirrels around the world.

FEBRUARY

_ . = • = ● = • =. _

Share your heart with citizen science and participate in **Beats Per Life** (page 53), a project that encourages you to sleuth in scientific literature for data on the heart rate and lifespan of a variety of animals. Your findings will help contribute to an understanding of how a long life may (or may not) be connected to heart rate.

Get ready to be a backyard ornithologist (bird scientist) during this month's annual **Great Backyard Bird Count** from the Cornell Lab of Ornithology. No equipment or experience required—you don't even really need a backyard.

The Great Backyard Bird Count (GBBC)

During this annual four-day February event (check project website for exact dates) bird-watchers count birds to create a real-time snapshot of where birds are located around the world, information from which scientists and bird enthusiasts can learn a lot. No single scientist or team of scientists could hope to document the complex distribution and movements of so many species in such a short time. Anyone from beginning bird-watchers to experts can participate in the GBBC. It takes as little as fifteen minutes on one day, or you can count for as long as you like during each day of the event.

Location Global

Materials Binoculars and field guides (optional)

Website Search "The Great Backyard Bird Count" on SciStarter or visit birdcount.org

Goal To create real-time snapshot of bird populations

Task Create an account on the project website or download the eBird mobile app, if desired. Count birds you see anywhere for at least fifteen minutes on one or more days of the GBBC. Submit a separate checklist for each new day, for each new location, or for the same location if you counted at a different time of day. Estimate the number of individuals of each species you saw during your count period and submit your results.

Outcomes More than 160,000 people across the world participate in the GBBC each year. In 2019, a whopping 224,781 participants recorded a record-setting 6,699 species.

Why we like this The GBBC is a great entry-level project that is fun for a family to do together and it helps the birds. Yearly data collection makes the information more meaningful and allows scientists to investigate far-reaching questions.

MARCH

_. = • = ● = • =. _

Have you ever seen the Milky Way? You're not alone if you haven't. Almost 80 percent of people never have. Light pollution from our cities and towns is a major culprit. You can measure the darkness of the night sky with **Globe at Night** (page 96). Simply step outside and mark down the stars you see to report the darkness of your neighborhood.

Chirp, chirp! You'll be hearing more and more bird sounds as spring arrives. Go through a short certification process and help track data on nesting birds near you with **NestWatch** (page 120). Find an active nest to monitor, report your data every three to four days, and help scientists understand the success and failures of nesting birds.

Rain, rain, go away . . . or stay so I can measure thee. Springtime brings rainy days for many—a perfect time for practicing citizen science with **CoCoRaHS** (page 84). You will need to purchase an approved rain gauge and set it up in a spot that you can check the rainfall every day. You might be surprised to find that rainfall amounts can differ drastically within a short distance. Your data will help provide a better picture of what is going on in your community. Another water-related project that begins this month is the **EarthEcho Water Challenge**.

EarthEcho Water Challenge

Water is key to life on Earth. Humans, plants, and animals cannot live without this life-giving ingredient. With the EarthEcho Water Challenge, all you need is a simple kit to collect and share data about water quality in your community. The challenge kicks off every March 22nd (United Nations World Water Day) and continues through December.

Location Global

Materials Water monitoring kit requested through the project

Website Visit worldwatermonitoringday.org

Goal To create a world map of the health of water bodies

Task Create an account on the project website and order your water testing kit. Read all the instructions to make sure you understand how to use it. Find a water body that you want to test and use your kit to collect the water sample and data. Submit your data online.

Outcomes The EarthEcho Water Challenge has a truly global reach—more than 1.5 million participants in 146 countries have monitored over 77,000 bodies of water.

Why we like this The EarthEcho Water Challenge is a great project for classrooms and community centers. The kits are relatively inexpensive and easy to use. Online action guides and lesson plans for several grade bands are available to help you implement the protocols.

APRIL

_.–•–•–•–.–

Save the date for **Citizen Science Day** in April. This annual world-wide event is organized by SciStarter and the Citizen Science Association to spread awareness of citizen science opportunities, celebrate the achievements of researchers and participants, and connect people to local opportunities. Events held during Citizen Science Day have included everything from bioblitzes, to hackathons, nature walks, and more. Citizen Science Day has grown in popularity and size every year since its 2016 founding. In 2019, librarians around the world hosted "megathons" in which thousands of people analyzed an entire year's worth of data for the Stall Catchers project in one hour! In 2020, the Earth Day Foundation and other partners will launch the world's largest citizen science initiative: Earth Challenge 2020. See the resources section for more information on Citizen Science Day and related programming.

When was the last time you looked for nature in your city? Another annual April event that encourages participation in citizen science is the **City Nature Challenge**. This challenge encourages people in cities around the world to find, record, and share observations of urban biodiversity.

City Nature Challenge

The City Nature Challenge was founded and is currently organized by the California Academy of Sciences and the Natural History Museum of Los Angeles. The challenge started in 2016 as part of the first annual Citizen Science Day and was a competition only between San Francisco and Los Angeles. During the week-long event, more than 1,000 people made 20,000 observations of 2,500 species.

In 2017, the challenge went national with sixteen cities and over 4,000 people participating to make 125,000 observations. And in 2018, the City Nature Challenge went global with more than sixty cities participating, including Hong Kong, Kuala Lumpur, Berlin, and Buenos Aires. Together 17,000 people made over 441,000 observations.

So what do people find in the cities around the world? It's not all squirrels, pigeons, and rats. Many people find endemic species—those that are only found in that particular city or region—and sometimes people get lucky and spot a rare or endangered species like an endangered woodpecker in Louisiana or a small flowering plant in the San Francisco Bay Area. Other people spot biodiversity that is just plain cool, like a falcon soaring around La Sagrada Familia in Barcelona, a bioluminescent fungi in Boston, or a chameleon changing colors in Colombia.

The City Nature Challenge continues to grow to more cities around the world. Learn more and find out how to get involved in your city at citynaturechallenge.org.

MAY

-.-•-●-.-.-

May is a great month to find a science festival. Communities across the world host big and small festivals during this time. Science festivals are great places to explore what your local community offers, interact with other science enthusiasts, and learn new things. Common events include scientist-led talks, expos, competitions, and more. If you'd like to find an event near you, check out the Science Festival Alliance at sciencefestivals.org.

Snapping selfies? You can contribute to the understanding of local waterways with **Stream Selfie** (page 91). Simply find a creek or stream, snap a photo (with or without you in it), and answer some quick questions. Surprisingly, many waterways in the United States are not mapped and rarely have important data collected about them. Your simple observations will help build a veritable picture of our water.

City dwellers unite: You are needed to help **Celebrate Urban Birds** (page 104). Learn the most common birds in your region (a great way to develop natural history skills), then find a place to birdwatch and commit to visit that spot for ten minutes a day for three days. Submit your data online to a database examining bird populations around the world. And don't forget to report back even if you didn't see any birds— that's important data as well.

JUNE

_. _ • _ ● _. • _. _

Hear that buzz in the air? It's pollinators. From bees, to bats, to moths, these important critters help bring us our food and our favorite flowers. Unfortunately, pollinators across the world are declining and scientists need help mapping their current locations. With the **Great Sunflower Project** (page 75), you can help by counting how many pollinators visit a Lemon Queen sunflower in a given time and recording other important details. Another way to get involved is to keep an eye out for bumblebees and contribute your sightings to **Bumble Bee Watch** (page 103).

Although scientists know that pollinators, especially bees, are declining, they aren't exactly sure what is causing the change. One concern is a small parasitic wasp that infects honeybees, causing them to exhibit strange behavior like leaving their hives and night and becoming attracted to lights like moths. With **ZomBee Watch** (page 98), you can keep an eye out for stranded zombees or build your own light trap to find them in your yard.

JULY

_. — • = ● — • — .

Enjoy spending time by the water? You can do that while contributing to citizen science by joining Surfrider's **Blue Water Task Force** (page 82) to test water quality along the shore. This important volunteer-led water quality data measures bacteria levels at marine and freshwater beaches. Data helps alert authorities to water quality issues along the coasts. The **Secchi Dip-In** (page 88), which runs for the month of July in concurrence with Lakes Appreciation Month, is another great project for monitoring lake water quality.

Sharks are important predators of the sea and with citizen science you can help scientists learn more about these creatures. If you see a shark while fishing or hanging out at the beach contribute your sighting to the **Shark Sightings Database** (page 90) to be used in research and understanding of shark populations.

National Moth Week (page 164) takes place during the last full week of July each year with events around the country that will teach you about this diverse group of insects. You can help scientists collect important data by contributing photos and locations of moths that you find through a variety of citizen science platforms.

National Moth Week

Moths—like butterflies, their daytime cousins—transport pollen between flowers and are vitally important to our ecosystem throughout their life-cycle. Their herbivores caterpillars munch on agricultural weeds, while adult moths serve as a food source for animals. There are more than 150,000 species of moths, ranging from the size of a pinpoint to the size of a human hand and spanning all colors. Scientists want to learn more about the distribution of these nighttime pollinators and they need your help. Enter National Moth Week, a project of the Friends of the East Brunswick Environmental Commission. Everyone is invited to observe, document, and map moths around the country.

Location Events are scheduled around the United States

Website Search "National Moth Week" on SciStarter or visit nationalmothweek.org

Goal To document moth species

Task Search the project website for public events near you or to find out how to participate on your own, such as finding moths near lights or during the day. Take a photograph of any moth you find and upload it to a National Moth Week partner, such as iNaturalist and the Encyclopedia of Life. You do not have to identify the species in order to submit photos. Include your location and other observations, following instructions provided by the app or website you choose.

Outcomes Hundreds of events take place each year resulting in thousands of new data contributions. Data is used to document the presence and absence of moth species around the country.

Why we like this This project can be done almost anywhere on any given night. It appeals to everyone from "moth-ers" (those who specialize in documenting moth species) to novices who simply share a picture for others to identify. Most people consider moths to be pests, but this project reminds us of the importance and beautiful diversity of moth species. You may never look at moths the same.

AUGUST

Summer heat is upon us in August and so are the biting bugs. Mosquito bites can be a simple nuisance during a summer outing, but they can also carry deadly diseases—mosquito-borne illnesses are estimated to kill 2.7 million people every year. The **Mosquito Habitat Mapper** (page 110) app will show you how to identify and eliminate mosquito breeding sites like old tires, flower pots, and other places prone to standing water. With simple equipment like macro-lenses for your phone you can also help identify mosquito larvae.

Celebrate National Dog Day on August 26th by completing a survey about your dog's temperament and behavior through **C-BARQ** (page 54). Not a dog lover? You can participate in a similar citizen science survey called **Fe-BARQ** for your feline friend.

SEPTEMBER

September brings the official start of autumn. This means weather, plants, temperatures, and rain may all begin to change. But, do you see change that is unusual? Something you've never seen before? Or something you've always seen and haven't seen this year? Report your weather-related weirding with **ISeeChange** (page 114) to help ground-truth the way changing data affects everyday lives. Your observation can be told as a photo or story and will be correlated with NASA satellite data from the same point in time that you made your observation. Your story helps scientists understand what the mathematical data points mean in real life.

Fall is also a perfect time to take a walk in the woods (or just down your street) and keep your eyes open for mushrooms of all shapes, sizes, and species. Submit your mushroom photos to **Mushroom Observer** (page 116); include identification if possible or ask for help from the project's mushroom community. By creating a place to share and discuss observations of mushrooms, the hope is to create a better understanding of mushrooms and ultimately increase our understanding of fungi.

OCTOBER

_.–•–●–.–._

What has eight legs and makes the average person cringe? With Halloween just around the corner, this is a great time to look for your spooky spider friends. You can submit your spider observations to **iNaturalist** (page 112) along with a photo, location, date, and time. Another ubiquitous sign of fall's arrival and decor are pumpkins, which are part of a family of plants found around the world and cultivated for agriculture. The **Great Pumpkin Project** (page 74) aims to document the microbe or insect communities that live alongside pumpkins and their relatives.

NOVEMBER

· — · • ▬ ● ▬ • — · —

Achoo! Do you have the flu? Winter brings cooler temperatures and the increased chance of feeling under the weather. Report your real-time sickness (or health) with **Flu Near You** (page 56). Simply answering the question "How are you feeling today?" can provide useful data for disease scientists who track flu outbreaks. Reporting your data weekly will provide a long-term snapshot of trends across the country. This data may help spot outbreaks before traditional methods do.

November is also the beginning of **Project FeederWatch** (page 72)— the project always starts in mid-November and runs until April. Counting birds at your backyard feeder is a great way to stay involved in citizen science throughout the winter.

Monarch butterflies make amazing migrations each year. In the Western United States, they travel south, sometimes even as far as Mexico, to over-winter. California is the only place in the United States where you can find overwintering monarchs, clustered together in the thousands in tree groves. Each year, volunteers with the **Western Monarch Thanksgiving Count** (page 125) collect data on the number, location, and type of monarchs they find. This data is critical in understanding how populations of monarchs have changed over time.

DECEMBER

_.·-•=●•-.·_

It's time to spice up your holidays with the **Christmas Bird Count**, which is considered the first and oldest citizen science project. The count actually began as a holiday tradition bird hunt, in which hunters would go out on their land with the objective to bring back as many birds and other animals as they could. But when the bird population decline became more apparent, this tradition was replaced with a bird census and the first modern count began on Christmas Day 1900. As you begin or continue your journey in citizen science, we hope you can take the chance to be a part of history and participate in this project.

Christmas Bird Count (CBC)

Everyone has their own winter and holiday traditions: meals with families, gift exchanges, travel, and, for some, citizen science. The Audubon Society's CBC takes place from December 14th to January 5th. Each winter, birders across the country get together to collect crucial data on bird populations. Find a location near you led by an experienced birder and get ready to participate. If you're new to birding, be sure to find an experienced birder who can assist with identifications.

Location At specific locations across North America

Materials Binoculars, field guides

Website Search "Christmas Bird Count" on SciStarter or visit audubon.org/conservation/science/christmas-bird-count

Goal To contribute to bird conservation

Task Find and join a CBC count compiler in your region for a count during the project's time period. You will follow a designated route within a fifteen-mile circle and count every bird you see or hear. You will be asked to follow specific protocols to make sure that the data collected around the world is consistent and comparable. The local compiler for your area will submit the data online.

Outcomes The CBC's long history gives researchers crucial long-term data about bird populations that isn't available through other scientific methods. The data, which has been used in climate change reports produced by the Audubon Society and the Environmental Protection Agency, shows important trends of bird populations decline over the years, bringing attention to species in peril.

Why we like this The CBC exemplifies the fact that citizen science projects collect data that could not be collected any other way. There is no way a single scientist, or even a group of scientists, could collect the number of bird observations over the project's hundred-year history. It takes people all across North America contributing data to collect this crucial data. The CBC is also structured so both novice and expert birders can contribute and work together.

RESOURCES
AND REFERENCES

Educators who want to integrate citizen science into their classroom or library will find support through resources and lessons plans available on SciStarter and other platforms like Students Discover, the Cornell Lab of Ornithology, NSTA's Science Scope written by SciStarter's Jill Nugent, and the Center for Community and Citizen Science at UC Davis.

※ Check out SciStarter educator's page to learn more about the research behind citizen science in the classroom: scistarter.org/educators.

※ Find lesson plans for teachers at studentsdiscover.org.

※ Visit the Cornell Lab of Ornithology's K–12 teacher online resources: birds.cornell.edu/k12.

※ At scistarter.org/library, find more tips and case studies of successful examples of citizen science in libraries in the downloadable *Librarian's Guide to Citizen Science*. You'll also find information about a series of citizen science kits designed for loan through libraries as well as how to use the embeddable SciStarter project finder widget, a user-friendly tool that will help you connect and engage your patrons with amazing research projects in your area and around the world.

※ Go to STARnetlibraries.org to find a wealth of STEM resources and programs for libraries.

Citizen Science Programming Ideas

Here are some of the many easy ways to celebrate the annual Citizen Science Day in April and/or promote citizen science in your community all year round.

✳ Find free resources, getting started guides, printed materials, events, and projects on scistarter.org/citizen-science-day.

✳ To collaborate on the development of open science tools, explore the Gathering for Open Science Hardware (GOSH), "a diverse, global community working to enhance the sharing of open, scientific technologies": openhardware.science/about.

✳ Host viewing parties of public television series for friends and family to learn more about citizen science. *The Crowd and The Cloud* and Nature's *American Spring LIVE* are both are geared toward an adult audience and the *SciGirls* citizen science series is mostly for young girls. These videos will provide context as to what citizen science is and how people can become involved.

✳ Organize a bioblitz: inaturalist.org/pages/bioblitz+guide.

✳ Advance community-driven solutions to local challenges related to natural resources, climate change, and natural hazards with support from the American Geophysical Union's Thriving Earth Exchange: thrivingearthexchange.org.

✳ Find information about collaborating with your community to improve public health and fight for environmental justice with Citizen Science Community Resources: csresources.org.

✳ Check out the California Academy of Science's citizen science toolkit: calacademy.org/educators/citizen-science-toolkit.

✳ Find a year-round catalog of citizen science events on SciStarter.org/calendar.

Further Reading About Citizen Science

Continue your citizen science education with these texts on your own or recommend a title to your book club to spread citizen science into the world.

Bat Count: A Citizen Science Story by Anna Forrester
Citizen Science: How Ordinary People Are Changing the Face of Discovery by Caren Cooper
Citizen Scientist: Searching for Heroes and Hope in an Age of Extinction by Mary Ellen Hannibal
Diary of a Citizen Scientist: Chasing Tiger Beetles and Other New Ways of Engaging the World by Sharman Apt Russell
Reinventing Discovery: The New Era of Networked Science by Michael Nielsen
The Rightful Place of Science: Citizen Science edited by Darlene Cavalier and Eric Kennedy

For more information on community-driven science, we encourage you to read this transcript of Dr. Max Liboiron's keynote address from the Citizen Science Association's 2019 conference: civiclaboratory.nl/2019/03/19/the-power-relations-of-citizen -science.

ACKNOWLEDGMENTS

We'd like to acknowledge the following people, without whom this book would not be possible:

First and foremost, all of the project leaders and citizen scientists who inspire us every day.

SciStarter's editorial and production team, each of whom made important contributions to this book: Daniel Arbuckle, Erica Chenoweth, Jill Nugent, Caroline Nickerson, Erica Prange, and Lea Shell.

Holli Kohl, NASA Goddard Space Flight Center, and Theresa Schwerin, Institute for Global Environmental Strategies.

Our agent Mackenzie Brady Watson, Stuart Krichevsky Literary Agency, Inc.

And our fabulous and talented editorial team at Timber Press: Mollie Firestone and Tom Fischer. Thank you for your guidance, humor, patience, focus, and unwavering support to help connect more people to opportunities to shape their future through citizen science. Thank you also to Julianna Johnson of Bologna Sandwich for the creative and intelligent illustrations that so wonderfully capture the essence of citizen science.

Darlene would also like to offer thanks for the support of Bob and her kids, Rebecca, Ronnie, Will, and Teddy, as well as Faith Quinn, Missi Vanover, Anne Goldman, and Sandee Cataldi. Catherine thanks Christopher Price for his patience and support and Dean and Linda Hoffman for encouraging her to write. Caren offers thanks for the support of Greg Sloan and the mini-Coopers, Abby and Zoe.

INDEX

monarch buttery, 15–16, 122,
125, 140, 168
Monarch Watch, 16
monk seal, 17
mosquitoes, 110–111, 165
Mosquito Habitat Mapper, 110–111,
165
Mother Carey's chicken, 28
moths, 162, 164
museum-based projects, 147–149
Mushroom Observer, 116–117, 166
mushrooms, 116–117

N

Nashville Zoo, 133
National Academy of Sciences, 14
National Aeronautics and Space
Administration (NASA), 108,
114, 132, 133, 166
National Audubon Society, 106, 170
National Moth Week, 163, 164
National Oceanic and Atmospheric
Administration (NOAA), 86
National Science Foundation, 42
National Water Quality Monitoring
Council, 81
Natural History Museum of Los Ange-
les County, 147, 160
Natural North Carolina, 148
nature journal, 112–113
Nature's Notebook, 18, 49, 118–119
NestWatch, 120, 157

night projects
Globe at Night, 96–97
ZomBee Watch, 98–99
nine-spotted ladybug, 29
noise pollution, 9
North Carolina Museum of Natural
Sciences, 147–148
North Carolina State University, 42,
53, 139

O

Oakland Museum of California, 149
online-only projects
Autoimmune Citizen Science,
64–65
Bat Detective, 52, 154
Beats Per Life, 53, 154
Canine Behavioral Assessment
and Research Questionnaire,
54, 165
DeepMoji, 49, 55, 154
Feline Behavioral Assessment and
Research Questionnaire, 54, 165
Flu Near You, 49, 56, 168
Foldit, 17, 27, 57, 154
Galaxy Zoo, 58, 154
SETI@home, 59, 154
Smithsonian Transcription Center,
60–61, 154
Stall Catchers, 154
open data, 40–41
Orbiting Carbon Observatory 2, 114

Darlene Cavalier's interest in science arose while working at *Discover Magazine* where she noticed that almost every kid-oriented STEM program had the overt message, "Learn this so you can grow up to become a scientist." But what if, she wondered, you didn't want to be a scientist? Where does that leave you in the world of science? Darlene went on to explore this question in graduate school at the University of Pennsylvania. She quickly learned about avenues to shape science and science policy and got hooked on citizen science. She created SciStarter in 2010 to help make citizen science opportunities discoverable to as many people as possible. Darlene is also the founder of the nonprofit Science Cheerleaders, which combines her love of cheerleading (she was an NBA cheerleader for the 76ers!) with a desire to make science accessible. The Science Cheerleaders—former and current NFL and NBA cheerleaders pursuing careers in science—work with fans and youth sports and cheerleading organizations to challenge stereotypes and engage people in citizen science.

Catherine Hoffman followed a mentor's recommendation to pursue citizen science as a way to combine her background in science with an interest in science engagement. After receiving her master's degree in zoology and relocating to a new city, Catherine ultimately found her way into citizen science via the online science community. She saw a tweet that SciStarter needed some help with project management on their database and jumped on the opportunity to learn more about the citizen science world. She has since served as managing director of SciStarter, started a sand crab monitoring program at an aquarium, attended conferences about citizen science, spoken to the public in citizen science trainings, and of course participated in projects. Some of her favorite experiences include collecting data for iNaturalist (she has surpassed 500 observations), swabbing her showerhead for microbes, taking backyard soil samples, and measuring the falling temperature in downtown Nashville during the total solar eclipse in 2017.

Caren Cooper's journey to citizen science began with parenthood. Although she didn't expect motherhood to change her career path—she had already been working and studying as a wildlife field biologist for more than a decade—field work lost its appeal when coupled with taking care of her family. Caren found her dream job doing research at the Cornell Lab of Ornithology, using data collected by tens of thousands of birdwatchers across the country to fuel research and conservation. She spent the next fourteen years at the Cornell Lab, with her interest in citizen science, and the citizen science volunteers, growing more intense with each consecutive year. When her undergraduate alma mater, North Carolina State University, started investing in citizen science she soon found herself back in her home state, mentoring graduate students to be public scientists carrying out their research in collaboration with citizen scientists. Crowd the Tap, a project from her lab, is featured in this book.